THE WAY HOME

No running water, no car, no electricity or any of the things it powers: the internet, phone, washing machine, radio or light bulb. Just a wooden cabin, on a smallholding, by the edge of a stand of spruce.

In this honest and lyrical account of a remarkable life without modern technology, Mark Boyle explores the hard-won joys of building a home with his bare hands, learning to make fire, collecting water from the spring, foraging and fishing.

What he finds is an elemental life, one governed by the rhythms of the sun and seasons, where life and death dance in a primal landscape of blood, wood, muck, water, and fire – much the same life we have lived for most of our time on earth. Revisiting it brings a deep insight into what it means to be human at a time when the boundaries between man and machine are blurring.

More praise for *The Way Home*

'[A] reflective, lyrical account . . . This genuine, warm-hearted analysis of the dysfunctions of our current world offers a surprisingly alluring alternative to our current malaise – if only we dared adopt it.'

Irish Times

'This memoir about living off the grid and tech-free in County Galway will inspire, connect and slow down the most impatient of readers, and that is a very good thing.'

Shelf Awareness

'A beautiful and thought-provoking story that will inspire you to live differently. Mark asks the most fundamental questions then sets out to live the answers.'

Lily Cole

'A revealing, humorous and deeply endearing witness statement on behalf of lovely, dirty reality.'

Jay Griffiths, author of *Wild: An Elemental Journey*

'A thought-provoking read which encourages the reader to appreciate many of the things we take for granted.'

Megan Hine, author of *Mind of a Survivor*

Mark Boyle is the author of *The Moneyless Man*, *The Moneyless Manifesto* and *Drinking Molotov Cocktails with Gandhi*, which have been translated into over twenty languages. A former business graduate, he lived entirely without money for three years. He has written columns for the *Guardian* and has irregularly contributed to international press, radio and television. He lives on a smallholding in Co. Galway, Ireland.

The Way Home

Tales from a Life
Without Technology

Mark Boyle

A Oneworld Book

First published by Oneworld Publications, 2019
This paperback edition published 2020
Reprinted, 2020, 2021 (twice), 2023

ISBN 978-1-78607-727-1
eISBN 978-1-78607-601-4

Typeset by Hewer Text UK Ltd, Edinburgh
Printed and bound in Great Britain by Clays Ltd, Elcograf S.p.A.

Oneworld Publications
10 Bloomsbury Street, London, WC1B 3SR, England

Stay up to date with the latest books,
special offers, and exclusive content from
Oneworld with our newsletter

Sign up on our website
oneworld-publications.com

For Kirsty Alston,
my mother, Marian Boyle, and my father, Josie Boyle

I am here not only to evade for a while the clamor and filth and confusion of the cultural apparatus but also to confront, immediately and directly if it's possible, the bare bones of existence, the elemental and fundamental, the bedrock which sustains us. I want to be able to look at and into a juniper tree, a piece of quartz, a vulture, a spider, and see it as it is in itself, devoid of all humanly ascribed qualities, anti-Kantian, even the categories of scientific description. To meet God or Medusa face to face, even if it means risking everything human in myself.

Edward Abbey, *Desert Solitaire* (1968)

Everything not saved will be lost.

Nintendo 'Quit Screen' message

Contents

Author's Note

Throughout this book I speak of places that are of special significance to me. But this is not a travel book, nor an encouragement to explore far-off lands that bear no relevance to your own everyday experience of life. Anything but. Instead it's an invitation to immerse yourself in your own landscape, to foster an intimate relationship with it, to come to depend upon it; to find your own place within your own place. This is work enough, believe me. As Patrick Kavanagh wrote in his essay 'The Parish and the Universe', 'To know fully even one field or one land is a lifetime's experience.'

Woven into these pages is the story of one such place, the Great Blasket Island, and the lusty people who scratched a living from its sandy soil and turbulent seas until their evacuation in 1953. As this tale of connection, loss and hope unfolds outside the book's seasonal rhythm, I have italicised those passages that step beyond the landscape around my neck of the woods, Knockmoyle, and enter into the lost world of 'Blasket time'.

Books tend to have the unfortunate habit of attracting thoughtless tourism to the places they reveal, the upshot of which can be the dilution of its essence and the particular things which made it worth writing about to begin with. If, for good reason, you still feel compelled to visit the places made known, all I ask is that you consider doing so in a way that their inhabitants, or the spirits that still haunt them, would welcome.

Places of character are full of characters, some of whom are human. All those I mention in this book are real, as are the stories and musings they imparted to me. To protect the privacy of my neighbours, however, I have given them fictitious names. On the off-chance that one of them ever stumbles upon a dusty copy of this book, I am sure they'll recognise themselves, and a few of the other characters, and chuckle. No one else need care; except, that is, for the names and characters of their own neighbours – human, and non-human, alike.

Prologue

I have written minutely of much that we did, for it was my wish
that somewhere there should be a memorial of it all, and I have
done my best to set down the character of the people about me
so that some record of us might live after us, for the likes of us
will never be again.

Tomás Ó Criomhthain, *The Islandman* (1937)

The afternoon before I was set to begin living in a cabin, without
electricity or any of the basic conveniences which, for most of my
life, I had taken for granted – a phone, computer, light bulbs, washing machine, running water, television, power tools, gas cooker,
radio – I received an email, perhaps the last I might ever receive,
from an editor at a publishing house. He had read an article I had
written for a newspaper, published earlier that day, and wanted to
know if I would consider writing a book about my experiences.

One year before that, when I first thought about building the
cabin – the bedrock for what I hoped would be a simpler way of life
– I came to the tough but realistic conclusion that, personal journals
aside, I would probably never write again. I was told that publishers
no longer accepted the hand-written manuscripts of D.H. Lawrence's
time, especially from people who were no D.H. Lawrence; therefore
my decision to start using less complex, more convivial tools was, I
believed, a death knell to the only financial livelihood I had. This I

accepted, as I had always maintained that, to borrow the words of
nineteenth-century writer and transcendentalist Henry David
Thoreau, it is more important to 'stand up to live' than to 'sit down to
write'. Still, the prospect weighed on my mind.

So his email came as a surprise. I told him that I was interested.
I had no idea at that point how it might work, if at all. For my
entire adult life I had used computers to write everything from
essays and theses to articles and books. I was already discovering
that hand-writing was not only an entirely different craft to
machine-writing, but that it involved a whole new way of thinking. There would no longer be the speedy convenience of the
typed word or online research, no spellcheck, no copy and paste
and no easy delete. If I needed to restructure a page, I would have
to start over again. I wondered how editing might work without
the instant communication that the modern publishing world has
become accustomed to. My mind boggled. There were a hundred
reasons why it might not work, so I picked up my pencil and set
about making that ninety-nine instead.

~

Almost a decade before I decided to unplug myself from industrial
civilisation, I began living without money for what was originally
intended to be a one-year experiment. It ended up lasting three
years, and money has played only a minor role in my life since. At
this point, you're probably thinking that here is someone with
acute masochistic tendencies. I could hardly blame you.

Strangely, the opposite is closer to the truth. Phrases like 'giving
up', 'living without' and 'quitting' are always in danger of sound-
ing sacrificial, limiting and austere, drawing attention to the loss
instead of to what might be gained. Alcoholics are more likely to

be described as 'giving up the booze' than 'gaining good health and relationships'. In my experience, loss and gain are an ongoing part of all our lives. Choices, whether we know it or not, are always being made. Throughout most of my life, for reasons that made perfect sense, I chose money and machines, unconsciously choosing to live without the things which they replaced. The question concerning each of us then, one we all too seldom ask ourselves, is what are we prepared to lose, and what do we want to gain, as we fumble our way through our short, precious lives?

As also happened with this book, the afternoon before I was due to begin living without money – living with nature still sounds too cheesy – I was asked if I was interested in writing a book about my experiences. One year later it, and I, would become known as *The Moneyless Man*. It was the story of all the challenges, lessons, miracles, struggles, joys, mistakes and adventures I had experienced during my first year of moneyless living. In the process of writing that book, my editor asked me to write a short chapter clarifying the 'rules of engagement'. As money is easily definable, the rules were straightforward: I couldn't spend or receive a single penny for at least a year. Considering my motivations were ecological, geopolitical and social as much as personal, I went to stupid lengths not to use the fruits of a global monetary system I was trying to live without. Ultimately, however, my self-imposed limitations were relatively clear and simple: no money.

So when the editor who first contacted me about the book you are now reading asked me to clarify the rules of my life without technology, it must have seemed a reasonable request, yet I instantly felt uneasy about it. Unlike money, it's not easy to draw a clear line in the sand in relation to what constitutes technology and what doesn't. Language, fire, a smartphone, an axe – even the pencil I write these words with – could all be described as technology, though I shy away from using such a rough brush to paint life. Where I would draw the line – the Stone Age? The Iron Age? The eighteenth century? – became an impossible question when the words themselves could be considered technology; and the more I reflected on my years without money, the less important finding the perfect answer seemed to become.

On top of that, those years taught me that rules have a tendency to set your life up as a game to win, a challenge to overcome, creating the kind of black-and-white scenarios our society leans towards. My life is my life, and it's prone to the same contradiction, complexity, compromise, confusion and conflict as the next person's. My ideals are often one step ahead of my ability to fully embody them, and that is no bad thing; in fact, as we will see later on, I wonder if hypocrisy might be the highest ideal of all.

I felt strongly that, if I were to write a book about my experiences, it ought to mirror what was the real point of unplugging: to deeply explore what it means to be human – in all its beautiful complexities, contradictions and confusions – when you strip away the distractions, the things that disconnect us from what is immediately around us.

Ten years on, I feel more drawn towards honestly exploring the complexities of simplicity, and less inclined towards being right. At the heart of how I live is the burning desire to discover what it might feel like to become a part of one's landscape, using only tools and technologies (if I must call them that) which, like the Old Order Amish people of North America, do not make me beholden to institutions and forces that have no regard for the principles and values on which I wish to live my life. And then, as life inevitably pulls me further afield – away from the hard-won simplicity of the cabin and the smallholding and into a society that seems to become more enthralled by virtual reality by the minute, to be free to recount the compromises and dilemmas I face, frankly and straightforwardly. Insofar as there are rules to my life, this is as much as I can say.

Within the limitations of words to accurately describe reality, the first chapter of this book intends to give you a flavour of the landscape which I am attempting to become a part of, and the cabin within which my new life began. The rest takes you through the seasons as I strip away the distractions whose convenience, I've come to believe, is killing us in more ways than one. Therefore the pages that follow are not so much the story of a man living without technology as they are a collection of observations, practicalities, conversations over farmyard gates, adventures and reflections, which I hope will provide an insight into the life of someone attempting to pare the extravagance of modernity back to the raw ingredients of life.

Actually, now that I think of it, this book has very little to do with me at all.

Knowing My Place

Would I a house for happiness erect,
Nature alone should be the architect.

Abraham Cowley, 'Horace to Fuscus Aristius. A Paraphrase
Upon the 10th Epistle of the First Book of Horace' (1668)

'This is the most beautiful place on earth,' remarked American writer Edward Abbey in his opening line of *Desert Solitaire*. For him that was the Canyonlands, the slickrock desert around Moab, Utah. But it was a title, Abbey knew himself, which had – and ought to have – no end of claims to it.

Such claims have been most vocal in the US. According to the poet and essayist Wendell Berry, heaven is Henry County in Kentucky, where he has farmed and stayed put while the rest of his generation, as Roger Deakin once put it, has been 'playing musical chairs' around him. There his tools of choice are a team of horses and a pencil. The conservationist Aldo Leopold probably felt the same about his shack on a sand farm in Wisconsin. To Henry David Thoreau, that place would have been Walden Pond for the two years, two months and two days he lived by its shore. For wilderness guardian John Muir, God's country was more expansive: the Sierras of the American West, from Alaska to the Yosemite Valley all the way to Mexico, where he searched out truths and challenged conventional wisdoms while 'carrying only a few crusts of bread, a tin cup, a small portion of tea, a notebook and a few scientific instruments'.

Over here, on my side of the Atlantic, Peig Sayers and Tomás Ó Criomhthain could have echoed Abbey's words on the Great Blasket Island, which is stranded 5 kilometres off Ireland's Dingle Peninsula and is the home of one of the most surprising, and forgotten, literary sub-genres of the early twentieth century. Over eighty books were written about or by these Blasket Islanders (though few are still in print), no small compliment considering only 150 people lived there in its heyday. Why the interest? Who knows. Perhaps intrigue, perhaps anthropological voyeurism or perhaps a sign of a generation who had lost something important and were told that it was last seen there.

To me, the most beautiful place on earth is this unsophisticated, half-wild three-acre smallholding in the middle of somewhere unimportant. It is here I wish to stake my own claim.

~

I landed on this smallholding in the summer of 2013, along with my girlfriend at the time, Jess, and a close friend called Tom. We were full of energy and bold, often unrealistic, ideas. After a decade living in England, the call to move back to Ireland was strong. I had missed my family, the people and the nuances of the culture. I had been away long enough for my Donegal accent to fade and for other Irish people to wonder where I was from. I was starting to wonder myself.

This was the first smallholding we had looked at. It was about as far from prime agricultural land as you could imagine, but it felt unpretentious, a place that was happy just to be itself. I remember, as we went to view the place, being struck by the gentleness of its atmosphere – the rustle of breeze on leaves, the hee-haw of a donkey, the coo of a dove – as we turned off the ignition in our camper van and walked up the track to where the potato field is now. The fallowness of the land seemed to me like it had important lessons to teach, lessons that might involve listening. We met a few of the more curious neighbours by the farm gate, and they were open, mischievous and warm. The air was alive with fresh manure, and we found it all strangely alluring.

As endearing as any of its qualities was the fact we could afford it. Ireland was in the aftermath of the 2008 financial meltdown, and we were offered the smallholding, and the farmhouse that came with it, for a rock-bottom price. My gain had been another man's loss. What could I do? Our budget was tight – stupidly tight – but I knew that having little or no money would mean that we'd have to get creative, and that this limitation could ultimately be our greatest ally.

We got to work immediately. We fixed up the house and converted living spaces into bedrooms, so that more people could live here. We planted trees, lots of them, while in other places we pollarded trees to let in light to the orchard and vegetable gardens we began to grow. The land was wet, so we dug out drains with our spades. We acquired a flock of hens, built a coop, planted a nuttery, created a pond, grew a herb garden, scoured car boot sales and junk yards for good quality, inexpensive hand tools that, for the sellers, were long-since obsolete. We made compost bins, composting toilets and, eventually, compost. We built a reciprocal-framed fire-hut that would quickly become a focal point for music, dancing and terrible hangovers. We only had a couple of months to get wood in and dry for winter. We scythed every blade of overgrown wildness in an unconscious attempt to put our own mark on the land, something I would later regret.

As I was also writing another book at the time, the workload took its toll on my relationship with Jess, which was already complicated by the fact that she wanted kids and I didn't, and we parted as good friends. I stayed here and she moved to County Cork. I promised myself I would never again put anything above a relationship, but I also knew that old habits die hard.

By the end of the first year I thought the hard work was done. What I have learned since is that the hard work is never done, especially when you reject all the things that fool you into thinking that self-reliant lives are meant to be simple.

~

I first met Kirsty in an enthralling, picturesque place called Schumacher College in Devon. Founded in 1990, it was named after the British economist E.F. Schumacher, best remembered for the classic book *Small is Beautiful*. She had been running a café at Alby Crafts and Gardens in Norfolk, where she was born and bred, but had slowly come to the conclusion that business wasn't adding anything to her own bottom line: happiness. Most of the time she found herself stressed, working every hour God sent and wondering what the hell she was doing it all for.

I was running a week-long course called 'Wild Economics' with a friend, the wild food forager Fergus Drennan, and she had come on it to explore other ways of making a living, ways that required little or no money. She had only decided to join the course at the last minute. It was a decision that was to have the most unexpected results.

We clicked instantly. I would find myself scanning the canteen, on breaks between sessions, looking to see if there was an empty seat beside her. We would stay up late talking, putting the world to rights. Her wide, deep brown eyes had a distinct sense of wonder that made you want to be in her company. We quickly fell in love. I once read that 'love is the recognition of beauty'. I saw many beautiful qualities in her – she was kind, playful, thoughtful, generous, she stood up for the people and things she cared about – that I had never encountered alongside such honesty before, and I felt blessed to have met her.

Within months we had begun creating a life together here. Neither of us had any idea how it was going to work. Kirsty was a wanderer who followed her heart, a dancer and performer who ran venues at festivals like Glastonbury. She had been wanting to live in a healthy relationship with the natural world, but had never before attempted to live directly from her immediate landscape and was uncertain about how she might find it. I was the stable, rooted sort who thought that mega-festivals like Glastonbury were an ecological travesty. But as every wild river needs solid banks, we felt that our differences could complement each other. Time would tell.

At that moment, all I did know was that I loved her, and that I would love her until my last breath, no matter how things would unfold.

~

Kirsty and I had been living in the farmhouse for almost a year when we decided to build the cabin. As I had previously lived in a 12-by-6-foot caravan for three years in England, living in a farmhouse felt luxurious at first. But I soon found that its conveniences

– switches, buttons, automation, sockets – were holding me back and discouraging me from learning the skills I wanted to learn and which I felt were an important part of the future, or mine at least. With electrically pumped running water on tap, I never bothered to walk to the spring.

In the farmhouse I found it difficult to look real life square in the eye, when electricity, fossil fuels and factories were taking care of it all for me. Having too much convenience is certainly a First World problem, but that doesn't make it any less of a problem, or one whose reverberations aren't felt in every nook and cranny of the planet. In the caravan I'd had a strong, direct relationship with the landscape around it, but now I felt like I was living vicariously through a seductive array of generic, functional gadgets. It occurred to me that perhaps the law of diminishing returns applied to comfort too, and that in the unceasing trade-off between comfort and the feeling of being fully alive, I was failing to find the right balance.

I wanted to feel alive again. Kirsty felt the same, though she articulated the urge to do so in her own way. We decided to let out the farmhouse, rent-free, to an eclectic collection of heretics – a yogi, two sailors, an anarchist, a circus performer and a musician – who wanted to live on the land, too. They all had their own reasons for wanting to be here, but a common thread connecting us was the feeling, understood in different ways, that something was deeply wrong with modern society, and that somehow we needed to reconnect with the natural world again, as much for our own sake as for nature's. This more collective approach to smallholding had been the vision for the place from day one.

With the cabin plans drawn up, the realities of the often-romanticised, so-called simple life were hitting home, the prospect of which aroused mixed feelings. We intended to be in and unplugged by winter, but first there was the small matter of building the cabin.

~

Sunday evening. It had taken me all week to dig out and level the foundations for the cabin. Twenty tonnes of hillside, shifted by spade. Just as I was clearing up for the night and thinking of a hot shower – might as well, while you still can, my body argued – a friend called over for a game of chess, his usual elaborate excuse for a glass of unusual wine (oak leaf on this occasion) and a chat. He said that he had heard that I was giving up technology, or something like that. Depends on what you mean by technology, I replied, but yeah, something like that.

He seemed genuinely concerned, not so much for me as for our friendship. How were we going to meet up? The same way we once did, I told him. Curious, he questioned me on the finer details – Email? Fridge? Internet access at the library? Clock? Running water? Gas? Public pay phones? Chainsaw? Wind-up radio? – to which I, in various ways, said no. As the conversation went on – it wasn't the first time I'd had it – he looked quite concerned for my welfare, too.

We've known each other since childhood, but there was a big gap in the middle, during which time we had taken different paths. He asked me why on earth I would do that to myself. Enjoy life, he said.

But that was the problem. I had stopped enjoying life. On one level I was enjoying blenders and toasters and once-unimaginable power, but I wasn't enjoying life.

I told him that I wanted to put my finger on the pulse of life again. I wanted to feel the elements in their enormity, to strip away the nonsense and lick the bare bones of existence clean. I wanted to know intimacy, friendship and community, and not just the things that pass for them. I wanted to search for truth to see if it existed and, if it didn't, to at least find something closer to my own. I wanted to feel cold and hunger and fear. I wanted to live, and not merely to exhibit the signs of life; and then, when the time came, to be ready to go off into the woods, calmly and clearly, and let the life there feed on my flesh and bones, just as I had done on theirs. Crows eating out my eyes, a fox gnawing at my face, a feral dog chewing on my bones, a pine marten making good use of my leg meat. It only seems fair.

While all of that was true, I kept the more important ecological, geopolitical, social and cultural reasons to myself. God knows, I could have offered up a few: the mass extinction of species; widespread surveillance in our bedrooms and pockets; resource wars; cultural imperialism; the standardisation of everything; the colonisation of wilderness and indigenous lands; the fragmentation of community; climate catastrophe; the automation of millions of jobs, and the inevitable inequality, unemployment and purposelessness that will ensue (providing fertile ground for demagogues to take control); the stark decline in mental health; the rise in industrial-scale illnesses such as cancer, heart disease, diabetes, depression, autoimmune diseases and obesity; the tyranny of fast-paced, relentless communication; or the addictiveness of the hollow excitement (films, pornography, TV, new products, celebrity gossip, dating websites and 24/7 news) that exists behind our screens, the goal of which seems to be the monetisation of our distraction. Etc.

But no one really wants to hear those – they're too preachy, too negative, too true – and so I poured us both another glass of wine instead.

After only a handful of moves we decided to abandon the game of chess, and re-corked the bottle of wine for another evening. He had to be up at 6 a.m. for work, and I had to start gathering round wood poles for the cabin.

~

It had been a tough-but-rewarding day. My feet were begging to get out of their boots, my back was glistening with salty sweat, my mind was clear and at peace. But the day had a bit to go yet. The sun was slowly cooling, so I took my neighbour's dog for a walk and went wandering in the woods, searching for the following day's building materials. I must have been gone a while, for both the light and my legs were beginning to fade, when I came upon exactly what I had been rummaging around for, every evening, for weeks. It was laid out so perfectly I wondered if it had been patiently waiting for me.

Stretching out 13 metres, the tree had, from what I could tell, blown over in the remorseless storms two winters earlier, its shallow roots unable to withstand one of the most tumultuous seasons Ireland had seen in many years. Having been suspended shoulder-height off the forest floor by a couple of neighbouring beeches, it was seasoning nicely.

Being a Sitka spruce, it was as straight as anything gets in the natural world, meaning it was tailor-made for what I needed: a roundwood ridge pole for the cabin, under which we hoped to spend our days. It was the combination of these qualities that persuaded me that it was finally time to round up an eight-strong crew of sylvan pallbearers to help bestow on it the respect that all life deserves.

That was the upside. The downside was that we had to get this cumbersome log out of the woods and down to where the cabin was growing up out of what grew around it. That involved carrying it up and down wet, boggy furrows for 300 metres, over an old stone wall, across a road and through an acre of copse to its final resting place. Even seasoned it weighed as heavy as an unkind remark, and all we had for the job were hands, shoulders, knees and pigheadedness.

But that was the deal.

~

Taking it slowly, half an arm's length at a time, Kirsty, a friend and I raised the ridge pole 4 metres into its new home, resting on top of a timber frame that would eventually give structure to lime-rendered, straw bale walls. Its sheer mass asked hard questions of every muscle, ligament and doubt in our bodies. As it should. We were told that some heavy machinery with forks could have done the job in half the time, and with just one person. But it felt good, important even, to raise this centrepiece together.

Over the following weeks, the roof took shape; sawn spruce boards with waney edges were overlapped above young, thinned-out spruce rafters, on top of which went the topsoil that we had dug out from the foundations a month earlier. Into this we broadcasted a wildflower-and-grass seed mix which, when they grew and blossomed, would blend the cabin gently into the landscape.

For all their beauty, wild roofs are not straightforward. One afternoon I climbed up onto the roof to fix a minor drainage problem. Looking around from that vantage point I could see the pattern of the place for the first time. It was a tightly woven fabric of people, wildlife, streams, fields, insects, trees, rocks and plants of which I was but one thread, no more or less important than the others. This place was no silk robe – it was more like the kind of Aran sweater that fishermen wear, but it felt hand-crafted, home-spun, rough around the edges and full of warmth. Sitting up there on the roof, surveying the landscape anew, I found a fresh sense of appreciation for this forgotten place in which I was slowly sending down roots.

Nestled in, as I was, among the dunnocks, bullfinches and robins perched in the alveoli of an old beech whose canopy partially sheltered the cabin's roof, my mind couldn't help but map the landscape around me. As I did so, I felt my own sense of self diminish within it.

To my right, as I faced the afternoon sun, lay our vegetable garden which, once our cabin-building was complete – would that ever happen? – was ready to provide us with roots, leaves, beans, courgettes and anything else that might do well in this soil and climate. When we first arrived here this patch had gone wild, or what is commonly called 'overgrown'. There had been a poly-tunnel on the spot below me, until we took it down and gave it to a friend, a market gardener who, despite thinking we had gone mad, was delighted to make use of it. I had two reasons for this. One, I no longer wanted to build dependencies on technologies whose manufacture I felt showed no respect for life. Two, both Kirsty and I wanted to live on an Irish diet – what this land can naturally provide without recourse to things like plastic – for better or for worse. We wouldn't be long finding out, but we were under no illusion that it would be easy, especially in times of great biospheric change.

It was this overgrown vegetable garden that had convinced me to move here in the first place. I remember the first moment I snuck my way into it through a shaded, secluded track arced by sweet chestnut, elder and hawthorn, where I came eye-to-eye with a handsomely fed stag who had been freely munching on wildflowers, grass and blackberries.

Considering that I ate a vegan diet at the time, and had always admired this fine beast's kin, I could not have known, as he stood there proudly in the reddish hue of an August sunset, that I would one day kill, skin and butcher some of his kind so that I, and the small woods I had planted, could live.

Right then I just stared in awe; of his form, his vitality, his gentleness and of that untameable look in his eye. It wouldn't be until a year later, when I would read Aldo Leopold's essay 'Thinking Like a Mountain' and attempt to live from these lands myself, that my thoughts on life and death would change dramatically.

~

On the opposite side of the cabin sits our hostel, The Happy Pig, which I built, during my second year here, out of local materials using natural building techniques, such as cob and cordwood, wattle and daub, roundwood and sawn spruce, and where visitors can stay for free. It is run in the spirit of a bothy, and it sometimes doubles or triples up as an event space and a *sibín* – a traditionally illicit pub made licit by the fact that any homebrew is served gratis. Through word of mouth alone it has become a halfway house, sanctuary or retreat for countless people who, for various reasons, long for reconnection with wilder places, or with the wild within themselves. We originally considered creating a website for it, but we were already too busy as it was, and I no longer wanted to go down that road anyway. There's got to be somewhere, after all, that isn't on the internet.

Head east past this hostel, through the nuttery and the potato field, and you come to the home of my nearest neighbour, Packie: a small white bungalow built in the 1950s to house rural bachelors. Packie is one of a dying breed of character, an endangered species who has that roguish glint in the eye you can only get away with when you hit your sixties. His face, like this place he never leaves, is well-weathered, showing all the signs of laughter and regret. What's left of his white hair is usually wild, except on Sundays, when you would barely recognise him.

I recall, on one of the first days after I moved here, having to write an article about 'gift culture' – a dry term anthropologists use to describe the myriad ways in which the first peoples organised themselves without money or barter – for a newspaper I've long since forgotten the name of. I spent all morning and afternoon working on the article, tapping plastic buttons to extol the many virtues of this natural form of economy.

The following morning, as I went to stretch my legs and get the lie of the land, I noticed that the field of grass I had scythed a few days earlier had been mysteriously cocked into neat mounds of hay. No one seemed to know how until the next day, when a rumour slowly got around that Packie, whom I had not properly met at that point, had been seen in our field, hay fork in hand, before the sun had even come up over his house.

He hadn't said a word to anyone.

~

To the south of the cabin grows a young orchard of apples, autumn olives, plums, sea buckthorn berries, pears, quinces, redcurrants and cherries, with some other useful plants like flax dotted here and there. Not all enjoy the poor-draining clay soil that defines life in these parts. Keep walking through this orchard and you eventually come to a quiet road – except, that is, at around 8:30 a.m. and 5:30 p.m. – which separates our smallholding from a stand of twenty-year-old spruce, itself bordered by a thin, deceptive belt of native broadleaves and an old thick stone wall of a pre-revolution, aristocratic estate.

This forest – perhaps 'tree farm' is a more apt term – was planted by man with timber and bottom lines in mind, yet it still has a distinct sense of wildness about it. It provides good homes for red squirrels, pine martens, hen harriers and wood mice (an important food source for native predators here). Just beyond my sightline were the places I would soon be foraging in with intent, skulking around picking sorrel, burdock root, chanterelles and wild raspberries. It is no cornucopia – few man-made forests in temperate climates are – but if you know where to look and what to look for, going for a walk with the dog can become part of your livelihood.

~

On all other sides, we're surrounded by fields of grass and tufts of dark green rush, the latter hinting at the clay beneath. This area is commonly considered to be marginal land, but I prefer to think of it as land misused by a society with marginal ecological understanding. Not so long ago it was a vast oakwood, the kind my generation finds it hard to even imagine today.

Derrybrien – a nearby village which, to the untrained eye, looks suspiciously like a pub – was anglicised from its old Irish name, *Daraidh Braoin*, meaning 'Brian's Oakwood'. It was here that a legendary High King of Ireland, Brian Boru, is said to have trained his men for some of his many battles with invading forces. As I sat on the roof I wondered what Boru would have thought of this modernising Ireland, and whether or not it was something he would still have given his life to defend.

All of this is connected by a network of narrow back roads – what we call *bóithríns* – with wild, unkempt verges and post-cardish strips of green grass running up the middle. Take the corner at the end of our *bóithrín* and you come to Kathleen and Jack's house. Kathleen is a small, hardy woman in her late sixties who possesses the enthusiasm for life of a six-year-old. She looks like she has popped out in 3D from a 1920s postcard, especially in winter when she wears a shawl around her head. Jack is older, in his eighties, and is now unable to do the things he used to – the neighbours have often told me 'he was some worker in his day' – which I notice sometimes frustrates him. Whenever someone from our smallholding goes down to help them bring in the turf or get Jack back on his feet after he has fallen over, he invariably tells us how he wishes he could do something for us in return. I remind him of all the times, when we first moved here, that he used his tractor to start our dodgy old camper van, and that every day I drink from the spring that emerges from their land. The conversation inevitably turns to Gaelic football, and all ridiculous notions of debt are forgotten.

You can't see or hear them from the roof, but beyond the hills that roll out in front of me are small parishes, villages and towns. The nearest shop, which doubles as a post office, is 6 kilometres away, and somewhere far off to the north-west is Galway City.

Down below, Packie was roaring at me to 'stop dossing and get back to work', so I climbed off the roof and came back down to earth, where a shovel and a pickaxe were waiting for me.

~

As I was embroiled in the frustratingly slow but important details of finishing a home, I noticed a couple of birds' nests nearby. The first was that of a swallow – the original cob builder – who had made a nest out of mud and straw in the roof of my shed. Who owned the deeds of the other nest I couldn't be sure, as it had recently been abandoned, but I felt a great sense of admiration for them nonetheless. Built with only beaks and claws – no power tools, no heavy machinery, not even a chisel or nail – it was made of broken twigs, soil, grasses, leaves and straw, and insulated with soft green mosses and liverworts, along with other materials from the landscape. While I marvelled at the determination and dexterity involved, I wondered if these birds' greatest skill might be their unending ability to keep their needs simple. Theirs is not a culture of progress, but one of artistic survival.

I went back to work, gently rubbing linseed oil into oak window seats and ash bookshelves, so as to strengthen and protect them. I paused briefly to consider what the swallow might think of my extravagance, before knocking such anthropomorphic nonsense out of my head. As I rubbed, the linseed oil transfigured the wood, bringing out its best, each ring and burr and knot telling stories about this place that had begun long before I was even born.

~

The cabin was finally finished. My body was sore. Sore and tired. Battling against the looming winter, I had worked every single day of the previous three months, most of which were spent carrying niggling injuries that had spread and multiplied through over-exertion. Every tonne of straw, stone, wood, earth and lime had been moved by hand, and it had taken its toll. The last week alone had been spent wheelbarrowing mud through mud as I dug out the drains – a job a more experienced builder would have done back in the dry, solid days of spring.

It was now winter solstice eve, the day before my favourite day of the year. Not only is the solstice Kirsty's birthday, but it also marks the end of darkness's dominance and the slow, grad-ual return of the light. Our ancestors, who didn't have light at the flick of a switch, celebrated it wildly, and for good reason. This year's winter solstice had added significance for me, it being the day I had set myself to start living without industrial-scale, complex technologies and instead to embark on a more hand-crafted path. I was planning to live in this way for a year at least, to see through each of the seasons in their turn, and to take stock of things once my opinions had been tempered by the experience. To hold strong views, either way, about the neces-sity or desirability of all of our machines seemed premature until then.

As I sat next to the fire, on the cusp of a new – yet much older – life, I could feel a sense of apprehension. Reality was kicking in. I was already feeling exhausted, and I hadn't even begun living without the ease of cheap fossil fuels and plastic switches and buttons. I was under no illusion that the way of life I was about to set out on was going to be some romantic, bucolic dream. Unplugging myself from the machine-world was about to shape my entire year ahead, and possibly the rest of my life. At this point

I had no idea if the insights it would offer me would be hard-won, or if I would take to it like a duck to water.

In some ways I'd been in a similar position before, at the start of my years living without money. But this time would be different. Very different. More primal, less insulated. For a start, I would have no solar panels, or the things they powered. No hot shower after a hard day's graft. No headphones or box sets or news or social media to distract me from myself. Experience told me that, as I stripped away the layers of over-civilisation like the skin of an onion, I would, in all likelihood, find out things about myself I hadn't known, and wished I hadn't known. Living far from the madding crowd, I wondered if I would feel a sense of isolation, or enjoy the peace and serenity. Having no internet, radio, television or easy connection to the outside world, would I quickly get bored? How was it going to affect my relationship with those around me, and my health? Was it even possible to live in an older way within the context of a modern society? The questions were plentiful, the answers yet to come, and I had a suspicion they would be covered in blood, sweat and . . . well, I hoped not tears.

To add pressure to an already challenging proposition, I had agreed to write a column for a newspaper that would explore both the reasons for unplugging and the actual experiences of doing it. I knew that even if I was able to thrive without technology I was going to get flak, but if I failed – something, as you'll see later, I have had experience of doing very publicly in the past – the criticism would be unforgiving. I didn't mind the thought of this for myself so much, as I have long since been used to it, but I felt uneasy about the thought of not doing justice to a way of life that had served my more competent ancestors well for millennia.

It was 11 p.m. when I checked my email for the last time, and turned off my phone for what I hoped would be forever. By unplugging myself from the wider, distant civilised world, would I lose all touch with reality, or finally discover it? I'd find that out too, soon enough.

Winter

... grace tangled in a rapture with violence.

Annie Dillard, *Pilgrim at Tinker Creek* (1974)

I wake up this morning to two thoughts.

The first is that, from this moment onwards, I haven't got a single bill to my name. I feel free. The second is that, from this moment onwards, all of the toll bridges linking my life to modernity are gone, and that I'm going to have to live on my gumption alone. I am cut loose from the only culture I've ever really known.

~

The first official day of winter, the solstice, has only just passed and already the finer details of this way of life are becoming apparent. At its heart and hearth is fire.

You never forget the moment you first make fire by friction. It feels primal, elemental, fundamental, essential. Any apocalyptic fears of economic collapse one might harbour would melt away at the vision of that primary, primeval incandescent coal, offering not only the promise of cooked food and warmth, but the reassuring knowledge that all is in hand.

I originally learned how to create fire in this way many years ago, but in the age of cheap and easy gas lighters I opted for convenience every time since and, in doing so, lost the most basic of crafts. Even when I was moneyless I would find half-empty lighters lying on the street, each one gradually eroding my motivation to keep alive something that, for reasons more important than producing an ember, ought never to be forgotten.

My story, in this respect, is a microcosm of those of a growing number of tribal peoples worldwide who, after coming into contact with the West and acquiring some of its tools, forgot how to create fire themselves. (Western 'aid' of sports t-shirts, jeans and trainers has similarly weakened their competence in making their own clothes and has, in effect, turned them into a new market for industrial clothing.) I remember once watching Ray Mears show a couple of tribal elders, who had lost their firecraft, how to depend on their own ancestral knowledge once again. Ray, as always, handled the situation with sensitivity, skill and grace.

Attempt one. Using a bow, I drill hazel into hazel all day, without so much as a coal to show for it. I know I'm doing something wrong, I've just no firm idea exactly what. What's worse, there's no one qualified to ask for guidance, or no instructive online video to watch. My head feels frustrated, my hips have seized up and my drilling arm feels like it's about to fall off. By the time I finally smell smoke, the only thing more painful than continuing to drill is the thought of having to start all over again. I give it my all, but my technique is lacking, and I'm spent. In more unforgiving circumstances, Kirsty and I would now be as good as dead.

Attempt two. I take more time preparing the wood. I whittle the drill and hearth to fit together neatly to increase friction, and select a more suitable branch for the bow. This one is longer, its arc slightly more pronounced, giving me greater friction with each stroke. Within thirty strokes it is smoking furiously, and what do I see but a magical, glowing ember looking back at me, quietly whispering, 'Good, you're beginning to know your place a little better now.' I transfer the ember, carefully, to a bundle of dead bracken and birch bark, blowing softly into it, drawing the flame upwards towards the heavens, towards God, towards food. Fire.

In this moment I feel like the world suddenly makes sense to me. Up until this moment, I just wanted to go to the shop and buy a bloody lighter.

~

Ever since my teens I've been a poor sleeper. I usually wake up at the first signs of the blue-black light of dawn. Having fought it for years with blackout blinds, the day I moved into the cabin I decided instead to embrace it, and to sleep with the rhythms of the seasons and my own body. Now the window above my bed, under the canopy of the old beech, has no curtains at all. I wake when I wake – no frustration, no expectation. But that's easier to say at the beginning of January than in the middle of June.

My mornings usually begin much like everyone else's: with the toilet. But that's probably where the comparison ends. If it is just a piss, I pick a tree and keep good my side of our symbiotic relationship. I feed it nitrogen, it feeds me oxygen.

Most mornings the situation is more serious, so I pay a visit to one of our composting toilets. I deliberately didn't build a toilet inside the cabin, which would have been easy to do. Modern houses are so well designed you almost never have to leave them. I wanted to spend as much time as I could outdoors, so I designed with that in mind. Know thy weakness.

My toilet is much like any other toilet, except that there's no flush. Or even water. But it has got a seat and is as comfortable as any I know. Instead of water we use either sawdust – which we collect by horse from a local backyard sawmill – or any other compostable material we have. Once the bucket is full I lift it out from under the seat and empty it onto the compost heap. Here it needs to decompose for at least a year, depending on the weather, but there's no hurry. The result is called 'humanure' and the plants and trees love it. Most humans don't. When you empty the bucket onto the heap, the smell can sometimes be rather pungent, and the first time can be quite an experience for many people. But like everything, it soon becomes normal and unremarkable.

Up in Dublin I'm told there are big protests about the water charges, which the International Monetary Fund (IMF) insisted on in return for credit in the aftermath of the banking crisis in 2008. It is the latest in a long line of new taxes which come at a time when people are having their homes repossessed by the very same banks the IMF bailed out. Up until then, water had always been free, at the point of service, to people living here. The government argued that it's very expensive to provide drinking water for millions of people, and that it needs to be conserved. The people argued that, until the greed of the banking industry came to light in such a brutal manner, this hadn't been an issue before.

I regret not being able to attend the protests, but I've no way to get there and plenty to do here.

~

If I had an FAQs page about this way of life, the first question I would have to put on it would be 'What do you do for Christmas without technology or money?' Everyone asks it – interviewers, friends, editors.

This Christmas Day? I got up early, fetched some manure, hauled logs from the woods for next winter's woodpile, and then remembered it was Christmas Day. We made food – roasted pota-toes, celeriac and swede, along with Brussels sprouts, salad and venison (after thirteen years of being either vegetarian or vegan, I'd decided to start eating meat – strictly wild, free-roaming crea-tures only) which a neighbour had dropped around – and drank some blackcurrant wine. I think we made love in front of the fire. Probably much like most people's day, but with logs and manure instead of phone calls and shit TV.

~

I find the pigeon in the verge. Roadkill. The thickness of its neck and head suggests it is a female. To say she was killed by a car would be true to some extent, but it would be about as accurate as saying that a mackerel was killed by a bottom-trawler, an oak by a chainsaw or a hilltop by a bulldozer; on closer inspection it could be said that they were all killed by an idea, one which is too busy to consider things like pigeons, mackerel, oaks and hills.

Whatever the roots of its demise, my eyes tell me that she has been dead for at least a day, while my nose informs me that she's probably still edible. First I cut the wings at the joints and the head by the neck, before plucking every feather out of her lean, supple body. The tail feathers are smeared in a yellow, watery, whipped raw egg-like shit, but I don't suspect she cares for ideas like dignity now. Looking at her dead naked body, I wonder if what I hold is still a pigeon at all. Is it the very essence of a pigeon, or literally everything but?

With my knife, I cut below the breastbone, and pull out the innards. As I do so, I realise that the pigeon and I have a lot more in common – a heart, liver, intestines, flesh, bones – than we have differences, most of which exist only as matters of arrangement and extent. I hope her spirit is soaring, but her body is staying here on earth, soon to take on a new form. The cycle continues. The cycle always continues.

I wash her out. She smells a little gamy, but she is fine to eat. There's not enough meat on her breasts to warrant taking her life, but more than enough to justify the little effort required to pluck and dress her. I stick her in the oven, and collect the feathers for some as-yet-undecided purpose. It will be a bittersweet moment if I collect enough to make a pillow.

~

A few years ago, before I rejected the internet, I was searching online for an image of a wild crab apple, hoping to make a positive identification. Instead of finding photographs of the plum-leaved or hawthorn-leaved crab, the screen was dominated by the trade-marked logo of the Apple corporation. Taken aback, I typed in 'blackberry' and 'orange' to see what would happen. I was offered mobile phone deals. I hadn't heard of Tinder at the time, but I don't imagine pictures of wood shavings, bracken and birch bark would have monopolised the page.

Six months later I read Robert Macfarlane's *Landmarks*, his remarkable, place-particularising contribution to a 'glossary of enchantment for the whole earth'. In it he revealed source of the words that had been deleted from the 2007 edition of the *Oxford Junior Dictionary*. They included:

> acorn, alder, ash, beech, bluebell, buttercup, catkin, conker, cowslip, cygnet, dandelion, fern, hazel, heather, heron, ivy, kingfisher, lark, mistletoe, nectar, newt, otter, pasture and willow.

In their place, Oxford University Press had added:

> attachment, block-graph, blog, broadband, bullet-point, celebrity, chatroom, committee, cut-and-paste, MP3 player and voicemail.

The publishing company's explanation – that these are the things that now comprise a child's life – was pragmatic, understandable, honest and deeply worrying.

In preparation for a life without the internet, a week or so before I unplugged I found a 2000 edition of the *Collins English Dictionary*; 1785 pages drawn from a 'Bank of English' consisting of examples of 323 million words. My own vocabulary has improved since getting it and using it to replace the online dictionaries I had used for years. If I wanted to understand the definition of a word in the past I would simply Google it, and by the time I had exhaled the 'w' of 'now' I'd have its meaning. But nothing else. Now if I want to find out the year Gerard Manley Hopkins died, my eye is caught by curiosities from hookworm (no thanks) to horn of plenty (another name for cornucopia – yes please) instead of a screenful of carefully targeted adverts.

Reading it is interesting. Only seven years older than the concise *Oxford Junior Dictionary*, there's no mention of block-graph, blog, bullet-point, chatroom or MP3 player. There's no entry either for currel – a word once specific to East Anglia which describes a specifically small stream – or smeuse, which Sussex farmers once called that 'gap in the base of a hedge made by the regular passage of a small animal'.

The smartphone generation, having never played with them, will not miss words like 'conkers'. It's odd – when I was growing up in 1990s Ireland on a working-class council estate on the edgelands of a struggling town, no one ever asked me if I missed anything about the natural world. But the moment I choose blue-bells over bullet-points I've found that everyone wants to know what I miss most about machines.

~

I'm trying to give up time. Obviously not seasonal time, the inescapable evolution-of-the-moment time; I mean clock-time. I appreciate that this may sound like a fanciful, impractical and odd thing to want to do, but it is at the heart of the way of life I want to lead. Reading Jay Griffiths' deep exploration of time, *Pip Pip*, reinforced in my mind how recent the concept of clock-time is in human culture, and how essentially ideological and political it is. Clock-time is central to industry, mass produc-tion, specialised division of labour, economies of scale and standardisation; basically everything I am trying to move away from.

That's the theory. But theory is one thing. Practically trying to extricate yourself from clock-time is something entirely different.

I have no watch, no phone, no clock. But I'm out chopping wood when the postman drives past. That means it's 9:10 a.m., more or less. My mind knows this. Packie strolls down the *bóith-rín*, takes a right turn in the direction of his sister's house where, every day – Saturdays excluded – he has lunch. That means it's 1:55 p.m. Previously I took great pride in knowing what clock-time it was from looking at the position of the sun, but now that knowledge comes back to haunt me. I have this inexplicable urge to spend just one day of my life seeing things as they really are, without reference to numbers or man-made concepts or anthropomorphisms or any human-ascribed qualities. Even one minute. And with that thought I realise just how far I still have to go.

~

Wash night. Sometimes it happens three times a week, sometimes only once, depending on what I'm doing.

It's a clear January night, the air outside cool with a razor-sharp breeze coming in from the north. I take off down the *bóithrín* with a couple of demijohns to collect water from the spring. It's a new moon, and I can barely see my nose, but my ears lead me to the source of flowing water, and the unmistakeable whirr of a bottle filling up tells me when the demijohns are full.

Back at the cabin, I light the fire, get the pot on the boil, and bring the bathtub in from outdoors, where it hangs on the spruce cladding. It's set in front of the fire, and into it goes a round wash bowl which is used to mix boiling water with cold. Depending on what body parts I'm washing, I'm either kneeling in the bathtub or hunkered over the wash bowl, splashing around or using a flannel.

It takes over an hour, and it's not a relaxing, soothing hot soak. It certainly isn't sexy or romantic. I've plans for a wood-fired hot tub outside, which has the potential to be both sexy and romantic, but for now, needs must.

Feeling fresh, I sit back in front of the fire and get out my book. Beside me, the cat licks herself all over before going out for a night wander, after which she'll no doubt be the perpetrator of crimes that no one will even notice tomorrow.

~

There is something quietly reassuring about a well-stacked woodshed. Henry David Thoreau, who went to live in the woods – because, as Lars Mytting noted, 'modern American society had become too hectic for him (that's right, in 1845)' – once wrote that 'after all our discoveries and inventions no man will go past a pile of wood'. I'm not convinced that this sentiment still holds as strong today, but I can certainly relate to it, and find myself prone to episodes of woodpile-envy as I walk past the harvest of another person's axe. A good woodpile suggests that I'm prepared. Therefore I'm never fully content until the following winter's firewood is in, ideally by the end of February.

The forest this morning feels calm and at peace with itself. All I can hear is a chorus of melodious birdsong – territorial claims, flirtations, warnings and conversations in birdspeak – along with the whinnying of an impatient horse off in the distance and the sound of my crosscut saw chipping its way through the years of a spruce. The air is full with the smell of citrus as my saw bursts bubbles of pine resin. A neighbouring robin stays in close attendance, surely expecting food.

As I cut through one log, the bark rips off the wood and exposes an entire tribe of woodlice. Their young fall into the dense jungle of the forest floor, while the old scurry around, their world ripped apart by a phenomenon so big they will never understand the nature of it. For them it is an apocalypse, their young dead or lost, their home destroyed, all broken. Whether it is an act of terrorism or a natural disaster seems to be of no interest to one woodlouse next to my foot. She is hell-bent on survival, looking for her young among the arboreal rubble.

~

A friend tells me that he once saw a 1970s German bumper sticker that said 'Everyone wants to go back to Eden, but no one wants to go on foot.' I want to walk. Eden doesn't exist, never has, but what is life if not the walk towards it?

~

I bump into a friend in a nearby village. I haven't seen him in a while, despite the fact that we live within cycling distance (18 kilometres) of each other and that he's one of my favourite people. We go for a pint, and after a few more than planned he tells me that he hasn't been calling over so much because of my views on the world. He disagrees with some, while others, he says, hold a mirror up to parts of himself that he feels guilty about. He says he doesn't mean for such things to get in the way, but they do. I had no idea, as I do my best to not talk politics anymore, especially over a pint. I assumed he had been busy. The thought that it may have been for ideological reasons never even entered my head. It's not an easy thought, but I'm glad I know.

As we finish our pints, I promise him that I'll call round and give him a hand in his garden within the next week, and he says he'll do the same. We both know how much more fun weeding and transplanting seedlings are when you do it with friends.

~

I arrive back from the woods for lunch to find a frozen deerskin lying outside my door. There's a note from Conor, a local carpenter, saying that he's never going to get around to using it and that there will be many more where that came from later in the year.

I've no fridge or freezer, and the skin is thawing out in this unusually mild January weather, so I spread it out on a 15-centimetre-thick spruce pole and get to work. Scraping flesh from the skin, out in the drizzling rain in the back of beyond, the twenty-two-year-old business graduate in me is wondering how the hell I arrived at this point in my life. To be fair, the thirty-seven-year-old vegan and animal rights activist in me is wondering precisely the same thing too.

~

A few weeks before I was set to reject industrial technologies, the *Guardian* got in touch to see if I would be interested in writing a column about the decision, and my experiences of doing so. I agreed, but I was aware that it would pose challenges that editors of international newspapers haven't had to deal with for a long, long time. The media world is now very much a digital one, built on speed, social media, twenty-four-hour news, devices, multimedia and all sorts of other things no longer at my disposal.

To my surprise, the editor there was accommodating and understanding. I get the sense he's even intrigued to see how it all works, and once worked. Our conversations went something like this:

– The articles are going to have to be hand-written.

– I hadn't thought about that. Okay, of course.

– I'll not be able to take photographs either . . .

– Oh, that could be a problem.

– . . . but my girlfriend would be happy to illustrate the column instead.

– Interesting, that could work nicely.

– I won't be able to comment on the online version of the articles.

– Again, I hadn't thought about that. We do like journalistic engagement, but it's not a problem. I'll choose a few representative comments from each article and forward them on to you when I reply.

I post him the first article, and with that my control over the process ends. I have to trust that he won't make any significant edits, as editors are often prone to do. I'll never get to read the article, online or in print. I'll never know how many people 'liked' it or shared it. Which is exactly the way it should be. For as soon as journalism becomes a popularity contest – rewarding sensationalism, groupthink and deceit over honest exploration of complex matters – people and places lose, and those who need to be held to account win. Win, that is, for a short-sighted moment.

Soon I receive a letter from the editor, along with hand-written letters people have sent in and a small sample of hand-picked comments from the online edition.

I start to read the comments pages. People are calling me all sorts of things, as predicted. Luddite. Smelly hippy. Middle-class, privileged white man. Misanthrope. Idiot. There are also a couple of thoughtful critiques, most of which I might have pre-empted if I hadn't been limited to twelve hundred words. I understand where all of the criticisms are coming from, as at various points in my life I could have written any of them myself, so I feel no ill will towards those who wrote them. Some I even agree with.

I open the letters. They come with real names and addresses, effort, and the unique style of the writer. There's a thoughtfulness about each one. Some are supportive, some are curious, some are critical. All are friendly. Most tell me that it's the first letter they've hand-written in years, and that they're really enjoying writing it.

The editor has asked me if I can respond to a selection of the comments. I think about it for a while, until Wendell Berry's words, in his poem 'A Standing Ground', come to my mind:

> Better than any argument is to rise at dawn
> and pick dew-wet red berries in a cup.

The time for red berries is still six months away, but there is cider
to be made.

~

When I managed an organic food company in Bristol, in England's
West Country, my life was full of keys. House keys, bike keys, a
whole other set of keys for work. I never really thought about it at
first, but as time passed they slowly came to bother me. I didn't
want to live in a place where everything had to be locked, and I
wondered why I had chosen to live among people whom I clearly
didn't trust. I sometimes thought about whether I owned the things
I locked up, or if they were slowly starting to own me. Yet out of
necessity – I had six bikes stolen in a three month period, five of
which had been locked – I would have to carry around a keyring of
jingling reminders of a way of life that I was beginning to doubt.

Ten years later, out for a walk, I realise that I've absolutely
nothing in my pockets. People sometimes suggest to me that I lock
the cabin while I'm out. I usually laugh and ask them to look
around and tell me what they think could be worth stealing. My
wooden mug? My carving knife? The artefacts which dwell in it
are only valuable to me.

Sometimes my old ways kick in and I consider getting myself a
bike lock, but don't. I know that something of greater value to me
would be lost if I did. By resisting that urge, I find myself spending
less time in places where I feel a lock would be appropriate.

A minor adjunct to this is that I no longer have to spend time
looking for keys. I wouldn't bother mentioning this at all if it weren't

for the techno-utopians of the transhumanism movement, who have taken a wholly different approach to the 'problem' of lost keys. According to Mark O'Connell, author of *To Be a Machine*, transhumanists – who, simply put, argue that our bodies are a 'suboptimal substrate' for our minds, which would be better off cased in machines (through mind uploading and other means) – cite the £250 million worth of time which British people annually 'waste' looking for keys as a sort of symbolic justification for what they call 'human enhancement' – words that wouldn't have been out of place in 1930s Germany – through such things as human implants and smart drugs.

That such a figure is derived from an assumption that every moment of our lives should have a financial valuation seems to be the least worrying aspect of the whole thing. The more worrying aspect is that many of the heads of Silicon Valley's Big Tech companies (some of whom, apparently, want be the heads-in-a-jar of Big Tech companies after they die) are powerful and wealthy transhumanists, and that they and others are pumping billions of dollars into a future for us as cyborgs. Those involved include PayPal co-founder and Facebook investor Peter Thiel, Google's Director of Engineering Ray Kurzweil and its former CEO Eric Schmidt, now technical advisor to its parent company Alphabet. The latter has said, 'Eventually, you'll have an implant, where if you just think about a fact, it will tell you the answer.'

I wonder who is going to be programming these implants, and what their answers will be.

Now that I think of it, though, we're only one step away from that point anyway. As O'Connell points out, our lives are 'increasingly under the influence of unseen algorithms, whose creators effectively control what version of the news we read, what we buy, what information we consume, even the romantic relationships we end up having. Schmidt's chip will simply mean we don't have to bother typing it into the search engine he runs any more. That way Google can know every single thing you think. Those who write the algorithms will rule the world. Perhaps they already do.

~

Packie has an old keg of Guinness at the side of his house. It has been here longer than me, and longer still. He has no idea how it got there, which considering it is entirely empty, I'm not surprised about. A few years ago he asked me if I had any use for it. I didn't, but Kirsty has just told me that she is missing having a hot cup of tea in the morning – no electric kettle or gas cooker – and so I decide to take him up on the offer and make a rocket stove out of it.

Making sure it is unpressurised, I cut a hole in its side and top and fix an elbowed flue pipe through its centre, finishing a few centimetres from the top. I insulate it with a natural material called vermiculite, which I found lying around in the shed, and with that we have an outdoor cooker for the times when it makes little sense to light a fire inside.

Rocket stoves are efficient. A handful of twigs is enough to boil a kettle, within minutes. The downside is that it has only one hob, which means that vegetables in the summer are usually eaten raw or steamed over a pot of potatoes. No bad thing.

I place the rocket stove on a cut of cordwood in our fire-hut, next to a box of dry twigs. The next day, at dawn, I see Kirsty wrapped up in woollens on a bench next to the new cooker, with a slight waft of smoke sifting up through the surrounding willow and shiny-green holly. There's an icy bite coming in on the north breeze, but she looks content, and I remember how much I admire her.

~

While I want my outdoors to be as wild and rugged as it wants to be, I prefer my indoors to be clean, tidy, calm and in order. Surprisingly, many wild animals are the same. I would go as far as to argue that most wild animals are entirely domestic in that they belong to a place; that is, their home. Domestication, in its truest sense, does not imply a lack of wildness. Domestication is an issue of control – of being controlled, and of trying to control others. One tends to follow the other, as we search for balance and attempt to wrestle back some semblance of control over our lives. I've yet to meet a truly wild person – a self-willed person not under the sway of mankind's opinions and its society – who has shown any desire to control others. The extent to which you are controlled by others is the extent of your domestication. It is the extent of your civilisation.

It's early, and Kirsty has the rocket stove roaring already, making the day's meal from the vegetable garden and the morning's tea – sage, lemon balm, chamomile, horsetail, mint, vervain – from herbs she dried last autumn. While the tea is brewing, I decide to clean the cabin. It's only one room, so nothing arduous. I sweep the floor with a wooden stick whose end is tied with a dried plant called broom (hence the alternative name for a brush). I clean the surfaces with water, and use the waste water for the house plants. I've not used sprays, detergents or even natural cleaners for over ten years, during which time I've also not seen a doctor. Some people use vinegar for cleaning – we make ours from apples – but I prefer to drink it and maintain my health from the inside out.

The cleaning is done just in time for tea. Kirsty is wrapped up in a blanket, and we sit together in silence.

~

After almost twenty-five years off, this morning I began fishing again from the shore of Lough Atorick, a nearby lake hidden from all but locals among a stand of spruce and expanses of bog. I was fishing for pike, but caught nothing; except, that is, for a small tyre which, for a split second, I thought was a monstrous fish. No such luck. Someone must have come to the conclusion that this idyllic lake was the best place to dump it, or the cheapest at least. I later find out that Lough Atorick is one of the few major lakes in the area without pike in it. The journey home is long. My friend Paul Kingsnorth, a writer and smallholder who lives a few kilometres up the road, tells me that I should write *The Vegan Guide to Fishing*, a step-by-step guide to always coming home empty-handed.

A local fisherman offers some advice. He says I'll need a boat if I am to have any joy in Lough Atorick. One of the couples in the farmhouse, Elise and Jorne, are experienced sailors and boat-builders. In 2006, Jorne set up a company called Fairtransport with two fellow captains in the Netherlands. Fairtransport ships cargo – including its own rum, coffee and chocolate, which been fairly traded with small producers – from the Caribbean to Europe, by sail only.

It's certainly not a get rich quick plan, as it's almost impossible to compete with the gargantuan cargo ships and the advantages which cheap fossil fuels and scale confer. To ship a bottle of rum with one of the behemoths costs roughly one pence. On one of Jorne's sailing boats it's more like one pound. And instead of taking days, it takes months. Despite these competitive disadvantages, some businesses are only too enthusiastic to ship with Fairtransport, so it continues to struggle and fight and survive.

But while I'm standing, hungry, by the shore of Lough Atorick, Elise and Jorne are back in the Dutch port of Den Helder, fixing up their houseboat, making it seaworthy to sail over here to the west coast of Ireland. That will take months, so any dreams I have of building a currach with them are on hold, for now at least.

~

I was born in the year the Pope came to Ireland, May 1979. I was given the names Mark Joseph John by my parents, and would later take the name Luke at my confirmation. This was Ireland pre-Father Ted. My surname, Boyle, is quite common in Ballyshannon, the oldest town in Ireland and the place where I grew up. A coastal town, its green rolling hills are separated from the Atlantic Ocean by miles of beach, home to some of the best surfing spots in the world, a fact few of us knew back then. In the 1980s we were glad to have a decent pair of shoes, let alone a surfboard.

My father, Josie Boyle, often told me that the River Erne, which flows through Ballyshannon into the Atlantic, was one of the best salmon rivers in Europe when he was a child. It was so wide it needed a fourteen-arch bridge. Against much local opposition, the Erne was dammed – and therefore damned – in 1952, to create hydro-electricity and jobs, or so the locals were told. After 1952 it needed only a one-arch bridge. Jonathan Bardon, author of *A History of Ireland in 250 Episodes*, recalled:

> Back in June 1944, John Gillespie, a native of Ballyshannon, had warned in the *Donegal Vindicator* that the 'mutilation of the river should be considered a national calamity'. His prediction that 'future generations of Ballyshannon people will not see a fairyland on their doorsteps' but that only a 'shrunken and imprisoned waterway will meet their gaze' proved all too true. The authors tell how the commercial

netsmen downstream suffered from the catastrophic decline in salmon runs, but they could have added more on the destruction of what Justice T.C. Kingsmill Moore described as the 'vanished Eden' – the 29 pools upstream which arguably provided the finest game angling in Western Europe and, at one time, employment for around 200 water-keepers and gillies and fly-tying business for Rogans of Ballyshannon.

We had no car growing up – which might partially explain why I still can't drive – so Dad would often take me on the bar of his bike to go fishing. But it was now much harder for him – or me, with my little telescopic kids' rod – to catch anything in this tamed, broken river. We'd catch the odd perch or small brown trout, but nothing that would feed a family. So, the times being as they were, we usually had fish only once a week, which my mother bought from the fishmongers, on a Friday. This would either be mackerel or salmon 'grown' in the salmon farm that the local authorities had established downstream from the dam. (A handful of workers now maintain the dam as a back-up for times when, like at half-time in the World Cup, everyone boils the kettle at once.)

I grew up on a street of eighty houses, including one which my dad was born in, and still lives there to this day. The only place he'll move to is the graveyard, and he'll put up a fight against moving there, too. As a seventy-two-year-old he cycled the 179-kilometre circumference of Ireland's largest lake, Lough Derg – itself 20 kilometres from our smallholding – in nine hours. He had what my generation would consider to be a hard upbringing. His father died when he was only twelve, after which he quit school and became the man of the house. Another education must have begun then, for when I was young I always admired how he could throw his hand at any job, and do it well. The same couldn't be said for me.

Until I was eight we all – Mum, Dad, my sister and me – lived with my grandmother, until she died in her sleep in the room next to me. I spent hours crying that day, though I wasn't exactly sure why, as the experience of death was new to me and it still hadn't sunk in that I would never see her alive again. Living as I do now, I wish she was still around, sitting on her chair by the fire, guiding and advising me on the best way to do many of the things she'd have known how to do as a child.

That street was where my friends and I would play ball games like 'kerbs' – standing on one pavement, you got points if you ricocheted the ball off your friend's kerb – until we were teenagers. Our parents wouldn't see us from one end of the day to the next. There were no mobile phones to check up on us, but every neighbour for a mile around had a quiet eye out for every kid.

There was only one phone among the eighty houses. It was located in the hallway of the neighbour directly opposite us. The door to that house was always open – as all the doors were then – and to use the phone you left twenty pence (the currency at the time was the Irish punt) on the table after your call. Phone calls were more expensive in the 1980s. I don't ever remember needing to use that phone, but I do recall my mother going in and out now and then, phoning relatives who had emigrated to England, Australia and Canada.

The last time I was back home I remember noticing how all of my neighbours' doors were shut. It felt strange, almost eerie. I remember watching kids I didn't recognise walk up and down the street, staring at their phones, scrolling up and down, oblivious to all else. They couldn't have played kerbs even if they had wanted to, as every square inch of pavement had a car half-parked on it. And I remember Gerry McDermott, a neighbour from across the street whom I've known since I was born, bursting out of his house to welcome me home, and to hand my parents a brown trout he had caught for our dinner.

~

It has only been four weeks since I gave up my phone, but it's strange not hearing my mum and dad's voices. Hard even. Harder for the fact that I know they miss it, too. They've supported me through thick and thin, and we've always been close.

This year there is no wishing them Happy New Year, no wishing my mum Happy Birthday. It feels selfish. Then I remember all the times over the years when I felt that my phone conversations were becoming a lazy substitute for going up to spend real time with them. They live 230 kilometres away.

I take out my pencil and paper. *Dear Mum & Dad.* I promise to come up to see them at least once every couple of months from the spring onwards, by which time I should have found my rhythm with this way of life.

Dear Mark. We understand. We love you too. Can't wait to see you soon. Take good care of yourself.

~

The potato field is a half an acre of poor-draining, muddy land, and it becomes obvious within weeks of moving in that we're going to need a good, solid path through it to the woodshed. My four criteria for any infrastructural work – natural, local, inexpensive or free, beautiful – apply to this job too, and that complicates things.

Gillis – a twenty-three year old from Flanders who showed up on a bike one day, as many do, to stay in our free hostel for a night and who, three months later, is still here – offers to help. I gladly accept. He's tall, broad-shouldered and strong, with a mop of blond hair and a powerful, gentle intellect which belies his age. He finds a small pile of flatstones on a patch of nearby commonage, cycles them back here, and that gets us started. But we need more. A lot more.

Seeing what we're up to, my neighbour Tommy Quinn drops in to have a look. Tommy's a sturdy character, a part-time farmer and part-time builder in his fifties, and the kind of man who would do anything for you. He tells us that he has a mound of megalithic stones that he had taken out of an old wall years ago, and that we are welcome to them if we can be bothered to get at them through the briars. We can. Options are limited.

We call in to Tommy's yard to see if they are suitable – flat, wide, straight-edged and 10 centimetres thick is just about ideal – and although most of them are much bigger than we would like, we decide that we have found our source. We need hundreds, but he has thousands.

Our first job is to get the best flatstones out of the mound and down to ours. Tommy's house and yard are a good 400 metres away to our west, and as some of these stones weigh over 30 kilogrammes, we wheelbarrow them across in threes and fours, more if they are smaller. Tommy tells us we're both mad, but I detect the slightest hint of respect in his voice as he says it.

Gillis starts on one section of the path, while I start on another. Each stone has to be offered up to the land for size, a hole dug out to its shape, and then played around with until it sits level with those that came before it. The cracks are filled with the excavated soil, which will one day sprout grass and make the path look as if it had somehow grown up there by itself.

The section of the path I'm working on is 15 metres long, and it will take me five days of heavy work. From my experience of using concrete, a day would have been more than enough to get the job done. But I'd rather work ten weeks at it than use bloody concrete.

Almost finished, I walk up and down it, checking for any loose stones. It feels solid. Looking at it in the reddish light of dusk, with the feeling of a good week's work in my body, I retire to the cabin content. That path will outlive me, I hope.

~

I call in to see Tommy's mother, Mrs Quinn, an eighty-seven year old neighbour from up the road. She is in great form for her age. For any age. She is a brilliant talker. Listening to her speak, as she sits huddled up to the range, you become acutely aware of how fast things have changed in rural Ireland over the last fifty years. We now live on the same *bóithrín*, but we grew up in vastly different Irelands.

The Ireland Mrs Quinn was born into was one before electricity, when the sky still had stars and people still had time for one another. It was an Ireland, Packie once told me, where people were afraid to put their first light bulb into its socket for fear of electrocution. She lived on one pence a week, eating cabbage, potatoes and other staples from the garden, along with eggs, pork and milk from their small menagerie. She earned that one pence per week from their eggs, and she usually spent it on wholemeal flour. For her, self-reliance wasn't a lifestyle choice. It was simply life.

These days my generation browse dating sites where everyone within the same country is a potential partner, and even national boundaries are no obstacle to love. Mrs Quinn tells me that, after she married her husband, it was a big deal to up sticks and move the 6 kilometres up the road, from where she was born and raised, to where she still lives today.

She could talk for hours, and I would happily listen, but it is past my bedtime. I promise to call round to see her again soon.

~

Having read many English translations of the books written by its last two generations of people, my first journey to the Great Blasket Island holds a sense of pilgrimage about it. My fascination with the place isn't for any of the reasons that tourists and day-trippers go there between June and August; instead, my interest is merely based on practicalities. The island folk of the Great Blasket were some of the last people that I know of to have lived in the way I am trying to live now, and I want to better understand how they did it – economically, culturally, practically. But I also want to understand why the Islanders of this remarkable place were evacuated in 1953, so I can anticipate any major issues.

The journey from our smallholding to Dingle – an old port town in West Kerry, which the Islanders would row into in their naomhóga *(canvas-covered wicker boats) to exchange fish for salt – would normally take four hours by car. Kirsty and I stick out our thumbs at daybreak, and my clock-mind estimates that it's roughly two hours before we get our first lift, having walked 8 kilometres down the road. There isn't a lot of traffic, and at various points we have to remind ourselves that those who look us in the eye and pass us are busy getting to work, running the kids to school, are unused to hitchers, cautious of strangers, or a hundred and one other things I don't know about. Eight and a half hours later we arrive in Dingle and, like the Islanders before us, we go for a pint of porter before making any big decisions.*

Dingle is throbbing with tourists like us. I remember reading James Rebanks' The Shepherd's Life *and being struck by his account of tourism in the Lake District in England – its population of forty-three thousand* residents have to cope with sixteen million visitors every year – where, in his words, 'the guests have taken over the guest house'. We're two more guests. As quaint and thriving as Dingle is, you get a similar sense here.

We have journeyed here without booking any accommodation – we haven't the capability to, even if we would have liked to – and have left it open to fate. I remember reading that, when the Islanders came in for their salt, the mainlanders would put them up for the night and offer them a horse and cart to get back to their naomhóga. We had decided to embark on this trip in these old ways, and hoped that we might meet someone over a pint and end up sleeping on a couch. After four years of running a free hostel, it's easy to forget that the rest of Western civilisation – even Ireland, which is slow in catching up – no longer operates in the same way. Our optimism for the old ways, on this occasion, has backfired and we now wish we had brought a tent. Still, it's good to keep that spirit alive, especially in times like these, when it is in serious danger of going the way of the dodo.

We ask around some youth hostels and are told it will cost €55 for a bed. Eventually we stumble upon a cheap room down one of Dingle's back lanes, throw our backpacks in, and go searching for a bit of traditional music that isn't amplified. We decide to make for the Great Blasket Island in the morning, and wonder what we will find there.

~

I decided to stop paying attention to the news in November 2015, over a year before I would reject the technologies that transmit it. It wasn't so much that I thought the news itself to be a bad thing per se – though almost all of it tends to be bad news – but more that I no longer wished to read it. I found it had become boring and repetitive. As Thoreau wrote in the nineteenth century, long before Twitter and twenty-four-hour news, 'If we read of one man robbed, or murdered, or killed by accident, or one house burned, or one vessel wrecked, or one steamboat blown up . . . we may never need read of another. One is enough.' The news had become a bit like a Hollywood movie – same storyline, different actors.

But no man, as they say, is an island, so the really big news stories found their way to my range of perception, even if only as bold headlines, whether I liked it or not. Trump. Brexit. The Syrian refugee crisis. Terrorism. Every now and then I'd overhear small talk of different people trying to be famous for fifteen minutes and not, as Pulitzer Prize-winning poet Gary Snyder recommended, for fifteen miles.

A few friends suggested to me that it's irresponsible not to keep up-to-date with global affairs, as otherwise politicians and big business will get away with murder. I get the logic, and perhaps they are right. But we've never been exposed to so much news, never had so many attentive followers of it, and yet politicians and big business are getting away with as much murder as ever. At the same time the ability of journalists to hold power to account has been eroded, as financially pushed editors favour quantity over quality to keep the Twitter feed rolling.

It's almost the end of January, and we've only had three or four wet days all winter. It's another clear, crisp morning, the white grass crunching beneath my feet as I start walking towards the post office. I call in to see a neighbour, and ask him if he needs anything, but it's really an excuse to see how he is. He's an old bachelor, living out here by himself, which can't be easy, and sometimes he gets a bit down – what our generation calls depressed. He says he's all right for everything, and we chinwag for a while.

Further down the road I notice that a horse has broken out. Until recently, if a horse broke out and a car hit it, the driver was responsible. The law has since been reversed, so that the owner

of the horse would now be responsible for any damage done to
the car. I go and find the owner, who is out fixing his tractor.
Together we walk up through the waterlogged fields to find the
mare and he gives me a short history lesson about the place.
Knockmoyle, he says, was anglicised from *An Cnoc Maol*, meaning the 'Bald Hill'. Looking around at the pasture here, I can see
how it got its name.

On the way back I find a dead fox in the middle of the road –
cars drive over and around it – and spot a pine marten shoot
across into a small wood, where it'll no doubt terrorise some creature that, at this moment, has no idea that this is the last day of its
life.

~

I'm reading Robert Colvile's *The Great Acceleration*, in which he
looks at how the world is getting faster by the ~~day~~ ~~hour~~ ~~minute~~
~~second~~ nanosecond. In it, he quotes an advertising slogan for the
BlackBerry Playbook. It goes: 'Anything worth doing is worth
doing faster.'

Good point, BlackBerry. Why spend an hour or two slowly
making love when you can fuck somebody for five minutes, after
all?

~

I haven't had a fridge or freezer for almost a decade. In a temperate climate, a metal box in a cool, shady place outside works just
as well as a white, electrified metal box in the kitchen inside, for
half the year at the very least. My decision to start eating wild
meat poses new challenges. But considering that humanity made

it as far as the twentieth century without an electric fridge or freezer, I'm quietly optimistic.

Fish are no problem. Usually they're relatively small, and so can be eaten by a small community of people over a day or two. I'd never take more than I needed. That said, some preservation for the winter months, when it is illegal to take trout and salmon (for good reason, considering how low their population levels are), would be very useful.

Venison, on the other hand, is a different matter. Kill one deer and you can get a lot of meat. The best place to store it is in the bellies of your neighbours, and Packie tells me that this was once the main method of preservation in Knockmoyle.

On the list of things I want to do on any given morning, killing a deer is second from bottom, just above buying plastic tubs of US-imported peanut butter. With that in mind, I know I need to preserve as much meat as I can from a single deer to get us through as much of each winter as possible.

Without a freezer, a smokehouse is the best option, once your neighbour's belly is full. Some people make them with concrete and sawn timber, so that they look like small sheds. Instead, I want to apply my four criteria again. So I take off to the woods for the morning with an irrepressible Gillis. Two hours later, we return with twelve young spruce poles, all of which were wind-fallen and cut with a handsaw.

Using a design from Ray Mears' *Outdoor Survival Handbook*, the smokehouse takes me two hours to make and costs nothing. It's big enough to smoke an entire deer. In time it will be covered by the skins of other deer, which local hunters have had no use for and have, up to now, been discarding. But as I currently only have one spare skin, I'll need to use a tarpaulin in the meantime.

It's dusk, and I still haven't taken Quincy, a dog I'm looking after, for a walk, so I bring her and Packie's dog Bulmers up through the woods for the final half hour of daylight. We're not long in, when I hear a rustle. I assume it's the dogs, who are off following their noses. Instead, out springs a stag, light brown and lean, his antlers wide and proud. He's beautiful. He stands in the middle of the track and we stare at each other, eye-to-eye, for what seems like a long moment. I wonder what kind of absurd beast I must seem to him, standing in my brown boots, muddy blue jeans and woolly jumper. And then in the next moment he's off, bounding through the young trees growing along the verges of the track. He knows where he's going. He knows every inch of this forest. He has to – his life depends on it.

Shortly afterwards the dogs re-emerge, looking none the wiser.

~

It has been over two months since I last used a phone, checked email or went online. As I write that sentence, I consider how much my life, and the world around me, has changed in the last twenty years that such a remark is even remotely noteworthy.

Sitting in candlelight, the soft light accentuating the grain of the beech table, I begin writing a letter to a friend in England whom I haven't seen since I returned to Ireland. I date the letter, and it takes me some time to move past 'Dear Emily'. When I used electronics I could average forty words per minute, but now . . . well, it doesn't seem to matter so much anymore. She had the thoughtfulness to write to me, so it will take as long as it takes to express the things I wish to say to her. I seal it, stamp it and put it in a small pile along with a handful of my other replies to this week's letters, something I do every Sunday evening. I'm not sure why I still see Sunday as the end of the week, but I do.

The following morning I walk the letters, along with the dog, the 12-kilometre round trip to the post office. The weather on the way is changeable – light rain one minute, heavy rain the next – and I feel invigorated by it. I arrive home to an unexpected and unknown visitor, who asks me how I have time for such slow travel in the modern world. I explain how, by getting rid of our van, I no longer have to work the two or three months it took me to pay for its purchase, insurance, tax, MOT, fuel and inevitable repairs, and that I certainly don't spend anything like that amount of time walking or cycling, which I enjoy doing anyway. She laughs at me, says I am completely mad, and we share a pot of chamomile and vervain tea.

The following day the postman calls in with a letter, and from the hand-written address on the envelope I can tell it is from Kirsty, who is visiting family and friends in Norfolk. I go inside to read it. It's what we, not so long ago, called a love letter, and I feel the same kind of youthful excitement I remember when we first met. I read it again in the evening, before putting it away, along with her others, in a drawer. She is travelling at the moment, so I can't reply to her, but I feel content just knowing she is out there in the world.

~

There's an old saying in Ireland that it is time to plant the potatoes when you can stand naked by your potato patch. For a hardy smallholder that could mean any time from early March onwards. Many would put them in the ground on St Patrick's Day, so that they would be blessed, but like any self-respecting Irishman I start thinking of planting the early spuds as soon as possible.

Packie's brother Mick, a softly spoken farmer who lives a short 300-metre walk up the *bóithrín* from our place, is in his yard sorting out the cows. He's getting on in years now, and it's starting to show, but if he stops he says he'll never get started again. I'd tend to agree, though his wife and hips seem to differ. He tells me that he has a mound of topsoil he doesn't need, and that I can take as much as I like. Tonnes of manure, too, he says. His yard is full of much else – old bikes, bathtubs, parts of things rotting and rusting. Mick, who like all of the remaining natives here is part of that generation which was taught to throw nothing out, grew up in a different Ireland, one in which people had next to nothing to hold onto. That wise old maxim has a different feel to it when applied in modern times. Mick loves his spuds, won't go a day without them, so I tell him I'll throw him a bag or two in the summer. He offers to drop the topsoil down in the bucket of his tractor, but I tell him I could do with stretching my legs, and go home to fetch the wheelbarrow. I have a feeling that my legs will be well stretched before the day is out.

There's a biting north breeze, but it's dry and fresh, with a sense of impending spring in the air. The landscape feels almost ready to burst enthusiastically into life, but the natural world is nothing if not patient. I barrow the topsoil first, as it will be laid first, shifting it the 300 metres from Mick's to ours, one barrow at a time. I barely scratch the surface, but the days are still short and the light has faded, so I'm ready to retire to the cabin and read by the fire.

This emerging potato patch – all being well, and all is not always well – ought to furnish us with around 150 kilogrammes of potatoes. Like the Stoics of ancient Greece and the Zen Buddhists of Japan, I try to be 'ablaze with indifference' to the fruits of my labour, but that can be easier said than done after a hard day's graft. Next month I will plant some oca, another type of tuber which makes for a blight-free alternative to the potato. I've not grown it before, so I am interested to see how it will do. Strength in diversity, and all that. We will see.

The day when I can stand by my patch naked is coming soon. I'm not sure, however, that the neighbours are ready for a time when my bare arse heralds the spring.

~

A well-respected environmentalist and friend of mine tells me that, though he understands my perspective on the ecological impacts of industrial-scale technology – basically that it depends on oil rigs, quarries, mines, the factory system, state armies, deforestation, urbanisation and suburbanisation, and everything else that environmentalists rail against – there's no way in the world he could give up his dishwasher. He's a talented, determined, adaptable man, so I ask him if he has suddenly developed confidence issues. But I do take his point – if he had to wash his own dishes by hand, where would he find the time to write and campaign against the ecological and social consequences of things like dishwashers?

I was an environmentalist once too, back in the days when it was still more about defending wild places and the natural world against untrammelled human ambition, and less about carbon and something obscure called 'sustainability'. It seemed to me, as I got older, that environmentalism was becoming preoccupied with taming these wild places – deserts, oceans, mountains – in order to harness green energy to fuel our way of life, and that of a small percentage of the world's people in particular. Paul Kingsnorth, in his essay collection *Confessions of a Recovering Environmentalist*, describes modern environmentalism as 'the catalytic converter on the silver SUV of the global economy', and suggests that it seems to be moved these days by a strange sort of equation: 'Destruction minus carbon equals sustainability.' So I gave up being an environmentalist – at least that kind – and moved out of the city and into the natural world.

My environmentalist friend and I have dinner. He's intrigued about how I'm going to wash the dishes, not particularly without the machine – he can remember how to do that – but without running water or washing-up liquid. I take him outside to where I store the wood ash, and mix a little of it with water to make a paste that will effectively be our washing-up liquid. It's an old camping trick, and it works excellently. If it were the spring we could have also used horsetail, a plant which invades our garden every year. I'd rather not have it at all, but we do. Having a use for it aids the weeding efforts. It is full of silica, meaning it is great for scouring pots and pans. As people need silica in their diets – it's important for hair, skin, nails and teeth – we also chop it up and put it into our salads.

Back inside, I fill a large bowl with spring water and get to work. Enthused by the novelty of it all, my friend wants to have a go at it. There's no objection from me, and I crack open a couple of bottles of homebrew instead.

After he's finished I pour the wood ash water from the bowl among a patch of birches I've planted. Ashes to ashes. Full circle. Life.

~

I awaken. It's pitch black outside as I go for a night-time piss. I don't know whether to stay up or go back to bed, and so I seek signs or suggestions of clock-time. Birdsong, the position of the moon or stars, a neighbour's light, anything. Nothing.

On a cloudy night, around this time of year, midnight and 5 a.m. can look much the same, especially if it is a new moon. I've no idea what time I went to sleep – I just remember it being dark and feeling tired – so for all I know it could be some ungodly hour of the night when I should be tucked up in bed. Or should I? If I feel rested when I awaken, why not just get up, regardless of the clock-time; or, if I feel tired or half-asleep, why not go back to bed? As I stand naked in the dense still darkness, I wonder what age I was when I began listening to a clock over my own body.

I come in from outside, take off my boots and put on some clothes. There's not a stir from Kirsty, who is a better sleeper than me. I light a candle and pick up my pencil. The time before the rest of the world wakes up is my favourite part of the day, and this morning that window of tranquillity has been extended. As I'm finishing my fourth page I hear a blackbird declare that his day has also begun, and I climb back into bed with Kirsty, who is slowly coming around in sync with the sun.

~

I don't remember a lot from my childhood. It was only the 1980s, but it feels like a different age as I reflect on it now. Looking back, we had very little by way of money, or the kind of things it buys, yet I've no recollection of ever having felt a sense of lack. I suppose we were all in the same boat, and in the days before the proliferation of aspirational television programmes, it wasn't so easy to feel the loss of a lifestyle neither you nor your ancestors had ever had.

It's funny the little, unremarkable things that stick with you through all those eventful and formative years. One of the most enduring memories I do have is of a steady flow of locals showing up at our front door with their bikes. I don't know whether it was because my father was passionate about cycling, or if it was just the normal way of things at the time, but I would later learn that he was fixing their bikes entirely for free. As a child I had no appreciation for what he was doing, or his reasons for doing it. In fact, as a child it seemed perfectly normal.

Life was soon to change. A beast called the Celtic Tiger economy was born and Ireland was, almost overnight, transformed from a financially poor nation to the sixth richest country (per capita) in the world. Investment was flowing in from the US, lured by tax breaks and a cheap seat in the European Union, which itself was throwing development money at us, mostly in an effort to modernise us, to turn us into an attractive market, and to make us as efficient as the more industrialised members like Germany, Great Britain and the Netherlands. Now that I think of it, it was around then that the stream of half-crocked bicycles arriving at our house dried up, too. I began hearing stories from my dad of people buying bicycles for the price of decent, second-hand cars.

It was obvious that a lot of people, especially in industries like construction and finance, had much more disposable income than before, but none of it was trickling down to the likes of us. As a thirteen-year-old I was working in a hotel, cleaning up vomit from the floor of its nightclub for an hourly rate of £1.50 (Irish punt), hoping to find the odd note or coin in among the empty beer bottles scattered everywhere. As a schoolkid, that was my way of trying to get some of the things flooding my world, things I had seen other people buying and suddenly wanted for myself. For those adults who couldn't work any harder than they already were, credit cards and loans bridged the gap between the new expectations of this tigerish Ireland and their daily reality.

It was with the scrapings of a nightclub floor that I bought a mobile phone, the first of my group of friends to do so. Considering I was also the first to get my hands on a Commodore 64 and a Game Boy, I was exhibiting all the signs of being an early adopter, and none of the signs of being an early rejecter. The phone was monstrous, and to get reception you had to pull a long antenna out of the top. My friends started calling me Del Boy, but I'm sure I looked more like Rodney. I'm not even sure why I bought one – except, probably, because I could – as in those days my friends and I would just call round to each other's houses unannounced, before going out to play Gaelic football. But within a few months, all of my mates had one.

We didn't know it at the time, but we were all enthusiastically taking part in the largest, most widespread social experiment in the history of human cultures, without any idea of its intended or unintended consequences.

~

The sky threw a real temper-tantrum last night. It was as if the gods had just found out about Formica, and wanted to punish humanity severely. For the first time since we moved in, the cabin took a real pounding, and I barely slept a wink all night. But it felt womb-like inside, and I arose this morning somehow feeling much more enlivened by the storm than tired.

Down by the spring, while fetching the water, Kathleen comes out to tell me that their electricity went out during the night, and she wants to find out if mine has too. She assumes I would know. It is too early in the day to get deep and meaningful about my technological choices, so I white-lie and tell her I have just woken up and have no idea. Before we know it a few of the neighbours, as usual, have congregated at the spring, and they're all saying that their electricity is out too.

As most people in these parts get their water electronically pumped up from their borehole and into a tank in their house, where it is often heated by a stove and pumped around its radiators, this means many will have to go without hot showers, hot water or heating until the local council fixes the problem. It could be hours, or days. Rural Ireland is never top of the authorities' list of priorities.

It's cold today, so I hope for Kathleen and Packie's sake that it comes on again soon. But I hope for the bowhead whale, the Arctic fox and the beluga's sake that it never comes on again.

~

Harley-Davidsons, SUVs and campervans are cantering along the winding, narrow road west of Dingle as Kirsty and I set off on the 20-kilometre walk to Dunquin. Its asphalted substratum has felt and supported the bare feet and hobnail boots of the Blasket Islanders countless times, and in a few moments of silent awareness I can picture them making their way home from the market, with salt on the back of a cart.

We give our thumbs the afternoon off as the journey is easily made on foot. I want to get a sense of what life felt like for a people whose fierce sense of place now seems antiquated to my mobile generation, and who would have made this long mountainous walk not as a one-off pilgrimage on a pleasant afternoon, but as a non-optional means of survival. My own experiences of living without a motorised vehicle, deep in the back country of rural Ireland, helps me to relate to how they may have felt in the rain and hail and darkness, and to see beyond – as best as one can – the rose-tinted glasses of an adventure.

As we ramble westwards, away from the hen and stag parties and towards the real hens and stags, the road slowly quietens. Drivers slow down and give us room as they pass, while some raise their hands from the steering wheel, a sort of muted hello to ramblers that speaks of something ancient and dormant within wayfarers of all kinds. We return the subtle salute, and there's a genuine feeling of warmth and thoughtfulness about it, despite driver and walker being separated by windscreens.

We stop at a pub called Páidí Ó Sé's – a shrine to a famous Gaelic footballer from the town of Ventry we're passing through – to rehydrate with a known diuretic: stout. It's what the Islanders would have done, if Páidí's had existed then, so who was I to argue? At the bar we meet a couple of local men who offer to take our backpacks to Kruger's Bar in Dunquin, where the Islanders would have once mourned loved ones and celebrated semi-arranged marriages on Shrove Tuesdays. They are off there themselves next, while they are still sober enough to drive. Kirsty gladly gives her pack – including her purse, identity cards and clothes – to the two strangers but, while I thank them, something stubborn inside me insists on carrying my own pack for the whole of the journey. We shake hands and promise to have a drink together in Kruger's.

My stout is almost empty when a phone goes off across the bar. It's one of the two men, phoning a friend to let us know that a thick fog has descended on the mountain pass. He's advising us to take a lift with another man at the bar, who he says will also be going our way. Before I know it the phone gets passed to me, and for a strange moment I feel compromised; I had sworn never to use a phone again, yet to refuse an act of thoughtfulness from a stranger on ideological grounds feels both wrong and absurd, and so for the first time in months I hear the electronic representation of the human voice. I thank him, and we walk the rest of the way anyway, fog or no fog.

He wasn't lying. We don't have torches, high-visibility vests or anything else people have come to regard as essential for an evening stroll, so we keep our senses alert and stand in the ditch whenever a rare car drives past on this mostly forgotten pass. It's not raining, but the fog is so thick that our clothes slowly become damp. I feel oddly invigorated. A woman stops to ask us if we want a lift, understands our objection, and drives on. Before dark we are having a pint with the men from Páidí's, one of whom initially pretends to be annoyed about us not heeding his advice.

Dotted along the walls of Kruger's I notice old pictures of some of those whose words I had devoured over the last year, yet only in that understated way that many pubs have photographs of their locals. Scattered among these are pictures of other locals who had acted in Hollywood movies such as Ryan's Daughter, Far and Away *and* Star Wars *– all of which were partially filmed in the area – but unlike Páidí Ó Sé's in Ventry, there isn't a sign of any of their biggest stars. I was told that Tom Cruise met Nicole Kidman there. Brendan Behan would come to visit old Kruger himself and get rat-arsed at the bar. Charlie Haughey – an ex-Taoiseach (the Irish Prime Minister) who bought one of the smaller Blasket Islands, Inis Mhicileáin, for peanuts – would regularly drink there, and while he was known nationally as one of the most corrupt politicians of recent times, the locals said he was 'great craic and much loved'. If all of these stories were true, they certainly weren't trying to cash in on them.*

We finish our pints in Kruger's, bid our farewells, and take off to find a place to lay our heads until sunrise. The island, as I gaze out at it through the back window, is shrouded in a thick, mysterious fog, and I can't help but hope that when I awake in the morning I don't find a place that, like the face of Che Guevara, has been conveniently wrapped up, packaged, sanitised, commercialised and sold to tourists like myself.

~

Back down at the spring, Kathleen tells me that her electricity is back on. Having grown up in the period before electricity had been rolled out across rural Ireland, a day without a washing machine or the television is no drama to her. At the same time, having long since designed a dependency on electricity into her home, not having it for a few weeks now would be a lot more difficult than when her family never had it at all. The dependency is not only practical, but psychological and emotional too.

We get to talking about the land where we live, which is partially covered in thick clumps of rush. She explains to me that this was never the case before, and that not so long ago we would have been looking out over fields of good grass (though itself a denuded landscape).

Theories abound as to why this might be the case. Some farmers say it is because the micro-climate here has been getting wetter in recent years. Kathleen suggests that it's down to the ways people like herself and her husband farm the land. Previously, farmers would take in the hay once a year; now they cut silage twice a year, a practice which ecologists know is responsible for the catastrophic decline in corncrake and curlew populations across the country. My own theory is that the weight of tractors is compacting the soil, something wet clay land like ours needs no encouragement to do. Whatever the reason, and it's probably a combination of all of the above, farming as a livelihood may, like the corncrake, no longer be viable in places like this one unless farmers start understanding the land as a biotic community. Who knows, that may be no bad thing. Perhaps rush is a pioneer species, spreading to force humanity out, allowing other species to return and slowly bring diversity and wildness back to the landscape.

By the time we have chewed the cud, my demijohns are overflowing. I bend down to drink directly from the spring. As it gushes into my mouth and over my beard – it has now been ten weeks since I last shaved – for a moment it is no longer obvious to me where the spring ends and I begin; the spring fills my mouth, flows south and slowly meanders into my veins, skin and bladder. Living as I do now, my health and the health of the spring are interdependent. If my neighbours spray insecticides and herbicides on their land, and they leach into the spring, I'll be poisoned with it, and so my fate and that of the wildlife here – the insects, fish, birds and mammals – are inextricably linked.

As I'm walking off, one of the farmers pulls up to fill his own water containers. It's a good sign. Great game at the weekend, he says, and I put my demijohns down and try my best to remember that there is absolutely no rush to do anything.

~

Tuesday night means Holohan's, a little pub perched quietly in an historic village called Abbey, 7 kilometres east of Knockmoyle. It's a glorious cycle on a clear night, and tonight is no exception. The sky is expansive, the Milky Way set among the constellations, the polestar standing loyally to my left, and there's not a single sight or sound of a car for the entire journey. It's good to be alive.

I park the bike outside, unlocked. Looking at it, I get the feeling that even it is becoming too fast for me, and that the journey would have been more magical on foot, and the silence more deafening. As I walk in, the landlord Tom nods at the stout tap and holds up two fingers, to which I give the thumbs up. The other pint is for Paul – or so I think – who lives on the other side of Abbey. Every week we meet in this halfway house for a chat and a game of chess.

We have one rule: unless your mum dies, or you break a leg, you show up, as there is no way of getting out of it last minute. As Paul still hasn't arrived by the time my pint is finished, I take out my book and get started on his too. He has no way of contacting me at short notice to cancel, and I have no way of finding out if everything is okay with him, so I hope his mum and legs are alive and kicking.

It's hard to read in rural pubs. The people are too friendly to let you get away with it. I overhear a woman at the bar telling one of the regulars how it is all changing for villages like this one. She has been watching the English news, she says, and has heard that two hundred pubs every year are going out of business over there. I know of five pubs within a 10-kilometre radius of our smallholding that have closed over the last decade. I'm not a big drinker by any stretch of the imagination, but I understand the importance of the pub to places like Abbey. Aside from the church on Sunday mornings, it's one of the few places where people still commune. When a village's pubs all close, any remaining young people leave for the cities. When the young people leave, the villages themselves slowly die, cottages go into ruin and the tractors get bigger. The woman at the bar says that pubs like this one will soon have to start putting on free taxis to entice people away from their cheap, supermarket bottle of wine at home.

As there is still no sign of Paul, I decide to call it a night. On my way out the door, Tom reminds me that there is a traditional music session next week. See you then, I tell him. One of the old boys passes me on his way in, and before he even makes it to the bar there's a whiskey waiting for him in front of the stool on which he always sits.

~

It's my first time sitting in a library for years. I sit by a desk, scribbling some notes from a few books I requested. Of the twenty-four people I can see among the various rooms around my desk, twenty-one are tapping on electronic devices – laptops, tablets, smartphones, desktops – while two others are reading newspapers. Only one person, other than me, is looking at a book. Even those desks that don't support an array of desktops have dual plug sockets mounted on their surface.

I climb the stairs to the first floor, both to stretch my legs and to search in vain for an old, rare classic. In among the aisles of paper I find myself alone. Looking over the balcony at the glare below, I watch a connected world grow evermore disconnected.

~

This is gorse country. It loves the place – the weather, the acidic soil and it even seems to thrive off the animosity it receives from its human neighbours. It has an undeservedly bad reputation, and I've never really understood why. Yes, it can spread quickly, but its bright yellow flowers – which smell of coconut on a sunny day – emblazon the otherwise green rolling hills every year between February and May. If parishioners here still made their own wine, attitudes towards it might change over the course of a single bottle produced from its flowers.

Up above the green and yellow hills falls a deep, infinite blue, and so it's the perfect day to go picking gorse flowers. Three of us, armed with buckets, stroll off towards a nearby lane which – because of its wild growing hedgerows and stone walls covered in all kinds of mosses, liverworts and lichens – the locals call 'the hairy *bóithrín*'. It's just about wide enough for a horse and cart, and has that strip of green grass up the middle that I find strangely comforting. We're determined to make 30 litres of gorse wine, so we're out to pick roughly 22 litres of flowers. Unlike dandelions or oak leaves, there's not a lot of substance to a gorse flower, so the thought of having to pick that many is a bit daunting as you start out.

The spines of the bush are prickly, and they help focus the mind. Try to work too fast and they are an excellent reminder to slow down. We know that it is going to take us all afternoon, and it's a repetitive job, so I try to absorb myself in the task at hand. My face is soaking up the sunshine, my nose full of coconut, my fingers tingling from the spines, my eyes marvelling at the peculiar and extravagant insects flying around me, each of them absorbed in their own work. We spend most of the time chatting and laughing, trying not to think too hard about how slowly the buckets are filling up.

Back at the ranch. Using dry twigs found scattered around us, we boil 30 litres of water in a big, blackened pot on the rocket stove. Once it is bubbling hot, we pour its contents into the bucket of flowers, where we'll leave it for a few days to stew. Job done, for now at least. We won't receive the pay for our afternoon's work until August, but the rate is good and the work pleasant. It'll be spent together just as it was earned together.

~

The day before I disconnected from the virtual world, I sent emails and text messages out to all of my contacts informing them of my postal address, as I had realised that most of my 'friends' would have no idea where I lived. With it I added a note, mentioning that if any of them were ever to show up unexpectedly on my doorstep they would receive a welcome that would have been normal for my grandparents.

A handful got back to me, before I unplugged, with an address of their own, and I stored these in a small, blue hardback address book. As this was my last link to these people, I made a paper back-up elsewhere. But some close friends hadn't replied to me by the time I signed off, and so on a wet Monday in March my mind wanders to Emily somewhere in Brighton, Mari somewhere in Finland, Adeline somewhere in France, Eric somewhere near Bristol, Markus somewhere unknown. I've no way of contacting them anymore, unless they get in touch with me first. That could be years, if ever. They could die and I may not even hear about it. It's a sad, unsettling feeling.

It's a Tuesday morning. A couple of friends show up on my doorstep, completely unexpectedly. They have come to visit for a few days, maybe a week. I'm busy, in the middle of a few seasonal jobs, and I feel an unwelcome sense of frustration come over me. Friends and strangers have been showing up fairly regularly since I sent out the message, making it quite difficult to get things done at times. But then I remember the tales my mother told me about her parents, who had more on their shoulders and less on their plates than me. We spend the day together, recounting the stories, adventures and struggles we each accumulated since we last met.

It's Wednesday morning. I start work before daybreak, hoping to get the essentials done before anyone else gets up. But I've no sooner got the trowel in my hand than my friends are up and out too. I'd forgotten they were early risers. They tell me they would love to get stuck in – they came to help out, they say – and so we chat and weed and shovel shit and laugh until hunger gets the better of us.

~

It has now been three months since I spoke to Mum and Dad. Their voices have probably been the only constants I've had throughout my life, but even those are now absent. I've slowly started to find my rhythm living in this way, so I decide that it's about time I travelled the 230 kilometres to see them. I've done a lot of hitching in the past, but when we bought a van five years ago its convenience dispirited my more adventurous side, in much the same way as charity shops have discouraged me from learning how to make my own clothes. I feel the urge for a hitching adventure once again, to start walking with no plan or expectations, and to open myself up to the inexplicable magic that often happens when you go in that spirit – or the tedious, depressing drudgery when you don't.

With small packs on our backs, Kirsty and I stroll down the *bóithrín* and, when we get to the small back road at the bottom, start hitching in the direction of civilisation – somewhere I haven't been for a few weeks. We've walked about 3 kilometres by the time the first car comes along. The driver stops to tell us she's only going as far as the next house, and we thank her anyway. Another half a kilometre down the road, a second car stops and takes us to our nearest village, Kylebrack, which sits on the main road to Loughrea, our nearest town.

This road is busy, at least as far as semi-rural roads in Ireland go. It's mostly used by commuters, passers-through going to work in Loughrea or, 35 kilometres further on, Galway City. Here we watch cars drive past our outstretched thumbs every thirty seconds or so – it's rush hour – and after what feels like half an hour we eventually get picked up.

On the N17 north of Galway City, the cars come in a seamless stream, yet we spend half the morning standing in the same spot. Gandalf would struggle to muster up the magic here. It turns out to be a big mistake going to the city. But once we get back out among the small towns and villages we've barely enough time to say goodbye and thanks to one driver before we're saying hello to another.

By the time we reach Ballyshannon we've had lifts from all sorts of people – a musician, a salesman, an ex-army engineer, a Sinn Féin politician, an accountant and a footballer. The only thing they all seem to have in common is that they spent years hitching themselves. One tells us he hasn't seen a hitcher for a long time, and we reminisce about our younger days when there could be a queue of six or seven hitchers on any one stretch of the road, all trying to get to work, visit relatives, or be anywhere other than where they were. At the age of twelve I was in that queue. Another driver brings us 5 kilometres out of his way, to a better hitching location, despite our etiquette-induced, semi-genuine protestations that there's absolutely no need.

Once, when I was living without money, I got a lift from a man (I think his name was Gerry) who told me he was fresh out of Portlaoise prison. Two years for assault. Portlaoise is where the hard bastards go. We travelled together for over an hour, and shared stories. His were more interesting than mine. I left a stainless steel water bottle in his car, but hadn't realised until we had gone our separate ways. The bottle would have been as good as worthless to anyone else, but as he knew I was living without money he understood its importance to me. Nothing to be done, so I carried on hitching. Forty-five minutes, and another two short hitches later, the old battered Ford Escort I had got out of earlier came screeching around the corner. Gerry had noticed that I'd left the bottle behind in the car and had been driving around trying to find me. I gave him an awkward hug, and we went off on different roads again.

It's mid-afternoon when our final lift drops us off in Ballyshannon, and Mum already has lunch made for us. She says she has been in and out of the doorway since midday to see if we had arrived. It is really special to see their faces, and we spend the evening by the fire catching up with all that's happened over the last three months.

~

After a week in which warm, dry weather fooled the daffodils, gorse flowers and me into thinking that spring had as good as arrived, a biting Siberian wind – dubbed 'The Beast from the East' by Fleet Street – makes it as far as Ireland, bringing with it an amount of snow not seen on this island in my lifetime. There are drifts the height of sheep resting against stone walls. The snow is pure, the type that demands to be transformed into snowmen and gently explosive projectiles. We oblige. It hides everything; except, that is, the usually secretive adventures of Mr Fox & Co., whose nightly rhythms have now been revealed for the less eagle-eyed like me. The deer have given their game away, and that may cost them dearly one day. Hunter-gatherers of old wouldn't have needed snow to understand the patterns of the fauna as I do now. They had fully functional senses, and a type of intelligence that has never been passed down to me.

There is not an event, big or small, on earth that is not advantageous to something, big or small, and therefore the question of good or bad is always a matter of opinion. To my eyes, the snow-bound smallholding appears magical, but as a Blasket Islander once said, 'You can't live off a good view alone', and so to the deer the situation may appear dire. To them, a late snow means that there is no food easily available at precisely the time of year that they have less fat on their bones to deal with the cold.

In Scandinavian countries, this would count as a mild, late winter's day. For Ireland, this is Armageddon. The neighbours tell me that, after two days of snow, the supermarkets in Dublin are being raided, one having had a bulldozer driven into it. We laugh, but wonder what might happen if climate chaos, as predicted, brings us weather patterns more catastrophic than a few days of heavy snow.

It being Tuesday night I set off, on foot, on the 7-kilometre road to Holohan's pub in Abbey. The snow is up to my knees in places. With the postal service out of action, I couldn't have cancelled even if I had wanted to. But I don't want to. With everyone else tucked up by the fire – or at least one of those gas heaters with the fake coals and flames – it means that I have the road to myself for the whole journey. I walk up the middle of it, along where the white line once was, through a glorious blizzard. By the time I make it to the pub, my beard has frozen stiff. Paul's there too, as surprised to see me as I am to see him. Besides the landlord, we're the only people there, and the spot by the fire looks even cosier for the walk.

Three days later, the postman is back on his round after an unexpected holiday. I receive a letter from a friend who tells me how, the day before 'The Beast', she met a friend in front of an empty bread section in a nearby supermarket. Her friend looked forlorn, so she offered to bake her a loaf. Her friend said no, she was looking for 'a sliced loaf'. My friend writes that she almost offered to loan her a knife, too. And while the bread section was decimated, she says the flour shelves in the baking section were full.

Another friend tells me that, at the peak of the snowfall, some people in Dublin were selling sliced loaves online for €100 each. I'm glad I'm not on the internet any more.

~

Trout season is open again. For the past six months they've been under the legal protection of the Irish state, the same entity which offers legal protection to the industrial agricultural practices that decimated trout populations to the extent that they now need protecting. Feeding your family with a trout caught at the end of January is no longer the basic human right it was before we invented basic human rights.

I revere trout – the creature itself, its spirit, its taste, and the vitality I feel when taking its flesh into my body on the rare occasion I get to eat it. There was a time when attempting to kill something that I claimed to love would have felt absurd, cognitively dissonant, grotesque, but now it seems quite natural. Don't ask me why, I don't know. But I'm not sure it's possible to truly love anything you don't depend upon. We only defend the things that we love, and you can't pretend to truly love anything you wouldn't willingly die defending. This perspective is similar to the widespread indigenous belief that in taking something's life – a plant or an animal – you take on a responsibility for the wellbeing of its 'tribe', and that you commit yourself to the defence of its species and its ecosystem, to the death if necessary. It's a perspective that recognises that the fate of each species, including our own, is linked to everything else.

I pack my rod and flies, hop on the bike and take off fishing for the rest of the afternoon. First I need to get some new line from the local tackle shop.

Looking back at my early twenties, I can see that I drew my sense of self-respect, to a large extent, from how much money I earned. We all did. As I stand in the fishing tackle shop waiting to buy a spool of 12-pound monofilament line – synthetic, disposable, cheap industrial stuff – I realise that I draw my sense of self-respect these days from how little money I need. This spring, once the nettles are out, I want to make my own fishing cord from its fibres, but until then I'm going to have to compromise.

From the tackle shop I head south-west, towards the river. My body is craving protein. Three months ago I withdrew my usual sources of it – chickpeas, peanut butter, tahini, butter beans, the usual international, industrial vegan classics – to replace them with something I now couldn't get a hook on. I haven't fished for trout in earnest since I was a child, and it shows. After three hours I haven't had had a single bite. Necessity, however, dictates that I persist. I don't know whether it is still too cold for trout, if I'm using the wrong flies, whether I would be better off using spinners, if I am fishing at the wrong depth or if I'm doing any of the twenty other things that I probably shouldn't be doing. I need to learn, and fast. I've enjoyed the afternoon's fishing – casting may be a more appropriate word – and while it certainly feeds the soul, it is not feeding the body.

One thing I have learned so far is that: it is futile learning how to fish without learning the river. Such things take time. I remind myself to have patience. It is on evenings like these that I wish I hadn't spent four years sitting inside lecture halls learning financial economics, when I could have been outside learning real economics.

~

I've just heard an interesting statistic: on average we touch our phones two and a half thousand times per day. Not having the internet, I'm unable to verify the source of that statistic, but my own observation of people in cities – sitting outside coffee shops, walking down the street, standing on the bus – gives me no reason to doubt it. A friend tells me that it is becoming common for people to use them in bed or while having a bath.

I light a couple of candles next to our bed. Tonight is massage night, something we try to make time for twice a week. I start with Kirsty's neck and slowly work my way down from the shoulders to her back, bum, hamstrings, calves and feet. The combination of being both a smallholder and a dancer means that Kirsty's lower back always needs particular attention. Being influenced by tantra, our massages – when our minds aren't wandering off somewhere else – are as much about conscious touch as they are about good, strong physical massage.

By the time it is over she is half-asleep, so I blow the candles out and we turn in for the night.

~

I was an average student. At primary school I showed some signs of promise, but after winning a scholarship to the local secondary school – which meant that I got all of my school books for free, a big thing back then – I lost interest quickly. Remarkably quickly.

It was 1992 and the modernisation of Ireland had begun. I had no conception of political forces and their agendas at the time, but on reflection I can now see how the curriculum mirrored the ideologies that were shaping this new, strange Ireland. In English we studied the poetry of Patrick Kavanagh, but not the spirited Kavanagh of 'Pegasus' which warns against selling your soul (which, throughout the poem, he describes as his horse), in a Faustian pact, to the devil:

> 'Soul,' I prayed,
> 'I have hawked you through the world
> Of Church and State and meanest trade.
> But this evening, halter off,

Never again will it go on.
On the south side of ditches
There is grazing of the sun.
No more haggling with the world . . .'

As I said these words he grew
Wings upon his back. Now I may ride him
Every land my imagination knew.

Instead we read the miserable Kavanagh, odes to dejection like 'Stony Grey Soil' in which the farmer-poet, speaking to rural Ireland, said, 'You burgled my bank of youth!' For some reason that line is one of the few things from school that has stuck with me to this day.

By the time I took my finals – the Leaving Certificate – at the age of seventeen, the only academic subjects I had any interest in were business and economics. It was in these that I got my top grades, despite – or perhaps, because of – being thrown out of both classes for challenging the teacher's views. Like everyone else, I had no idea what I wanted to do with my life, but like everyone else I was told I had to do something. I got offered a place on an undergraduate IT course in Belfast, which was where my five best friends had agreed to go, but at the last minute I decided to accept an offer to study business and economics in Galway instead. At that moment I didn't know why I had made that choice, and I certainly couldn't have foreseen the ways it would unfold in my life thereafter. I had just followed a hunch.

~

It's the second Tuesday of the month, and that means a trad session in Holohan's. I'm here with Gillis, breaking up the cycle ride home from the lake. It's supposed to start at 9:30 p.m., but there's a healthy contempt for precision in places like this. The small hand of the clock on the wall is touching ten when one of the regulars walks in. He sits in the corner saying very little most nights, but tonight he has got the squeezebox under his arm, and as he walks in everyone in the pub looks towards him. In quick succession all eight musicians – guitarists, fiddlers, tin whistle and *bodhrán* players, and singers – are in, with their drink of choice in front of them without so much as a word. All of the musicians glance towards the squeezebox. He takes a sip from his whiskey, gives the nod, and off they go. For the next three hours eight musicians will weave their arts into one long, captivating, melodious whole, governed only by custom, tradition and the timeless uncertainty of emerging genius.

Not long after they begin, two young girls and their mother – relatives of the landlord – walk in, full of smiles, arms loaded with tin-foiled plates. They shuffle quietly out the back, before re-emerging shortly afterwards with a platter of small, triangular sandwiches. To say that these sandwiches were free would be a denigration of the spirit of the act; the thought of charging for them doesn't even enter their heads, it's just what you do. Gillis and I are famished, and so our eyes light up at the sight of it. We have more than our fair share, but as Gillis has never been to a trad session before and is clearly enjoying it, we seem to be having more than our fair share of pints too. We only came in for one, or so we claim.

After every few reels one of the musicians, or someone at the bar, pipes up with a song. At these points the whole of the pub goes quiet, and anyone daring to talk quickly gets shushed. The songs are sometimes lilting, other times reminiscent and sad, and are always followed by a hearty round of applause.

Just as the musicians are finishing up we head for the door, and ready ourselves for the final leg of the journey, all the way uphill to our beds.

~

Lunchtime. Rain is in the air, I can feel it, and as I want to get the rest of the wood in under cover while it is still dry, I'm not keen to stop too long to eat.

I grab a small willow basket and scurry off around the land looking for salad. It's a good time of year for wild food. Dandelions (flowers and leaves) and ground elder – both of which are considered weeds and are distinctly flavoursome – are already out, and I mix them together with ramsons, wood sorrel and navelwort. We're just coming out of 'the hungry gap' – that tough period in places like Britain and Ireland when the winter brassicas are finishing and the stores of roots are running out – and we've almost nothing in the garden, so this new growth of greens is welcome.

A few leaves and flowers won't keep me going until evening, however, so I take three hens' eggs, crack them on the side of the mug, and swallow them raw, straight down the hatch. A bowl of oats and I'm off out the door again.

It's pissing it down as I haul the last log of the day through the copse and into the dry of the lean-to, each step squelched through the March mud. I was already tired and hungry and dirty, and now I'm soaked to the bone.

Despite knowing little or nothing of the bloody, mucky realities of land-based lives, people sometimes tell me to be careful not to romanticise the past. On this, I agree. But I tell them to be even more careful of romanticising the future.

~

In times like these, when scale and efficiency mean everything, a place can change overnight. I hadn't been for a walk in almost a week, but I had heard a clatter of machines working in a nearby forest during that time, and I guess a part of me didn't want to look for fear of what I might find.

I decide to look. I turn the corner by the old gatehouse cottage, which is being renovated to rent out to tourists on Airbnb. My fears are confirmed. The forest, in which I've walked dogs most days since moving here, is gone. Just like that.

Coillte – the semi-state body responsible for managing many of the country's forests – are now clear-felling areas that had become part of my life. It's not that you don't know that it's going to happen. You do. It's a tree farm, after all – well, to them at least – not a self-willed woodland, and despite its name (*coillte* means 'woods' in Irish), they are in the business of producing timber, not protecting woodlands. All things considered, there was no reason for me to be shocked when I saw it. But I was. To the wildlife – fallow deer, woodlice, pine martens, pygmy shrews, red squirrels, midges – it must feel like the equivalent of the A-bomb on Hiroshima.

It's the speed that is the most startling. One minute it is there, the next minute it is gone. The machines which do the dirty work look like something out of the film *Avatar*. They do everything: fell the tree, lop its side branches, lift it, move it, stack it, all within the blink of an eye. Human feet don't even need to touch the forest floor. These machines have none of the stiff rigidity of diggers, but all the dexterity of the human arm and hand. I watch one grab what looks to be a twenty-five-year-old spruce; it tosses it up like a toothpick, catches it, lifts it onto a trailer, all in one fell swoop. Then onto the next one without having to catch breath. Deadly efficient. There is already a bank of hundreds, perhaps thousands, of spruce trees neatly piled alongside the track where I walk

Bulmers. The machines are using the collateral damage – the native trees, along with the invasive rhododendron and laurel that thrive in such acidic environments – as structure for their caterpillar tracks to move efficiently over.

After staring at them making cricket stumps out of 10-metre trees for a short while, I walk on. Bulmers seems more interested in a smell he has stumbled upon in the long grass. I know that in a short time I'll get used to this war zone, and I think that's what bothers me most. I know that all of the mammals will have got out in good time and that, for now at least, they should find other places to go. Their habitat is becoming more scarce, however, and the herd will always be made to fit the range. For those other non-human tribes we no longer recognise or pay attention to, it's the end of the world as they know it.

~

Kirsty and I have had a tough day, physically and emotionally. Coming off any addictive drug is hard, and technology is no different. It doesn't happen very often, but we both wish we could just chill out in front of a film for the evening and get out of our heads. That's not an option anymore.

It's dark. I light a couple of candles, and we lie in each other's arms in front of the fire. We talk now and again, but are mostly silent. Sleep will come, and tomorrow will be a different day, when things will look and feel a lot different. I knew we would have moments like this. I also know that such moments will become rarer as time goes on and the temptations of technology weaken their hold on us.

The fire burns orange and red, and soothes us to sleep.

~

It's morning, and the dense fog around Dunquin has lifted. Walking down the hill towards the Blasket Centre – a museum of the islands – I think to myself how minimal and unobtrusive the marketing of the centre, and the islands themselves, is around Dunquin. Understated on the outside, its insides are quietly impressive but, unlike the islands' houses, which thus far I've only seen in old photographs, it is similar to all museums: oversized, spacious, grandiose.

I wander around it for a couple of hours. I take notes on the tools they used. I examine the innards of a mock-up set of what their stone cottages would have been like, and read about academics like Robin Flower – whom they affection- ately nicknamed 'Bláithín', meaning 'little flower' – and George Thomson, both of whom encouraged and helped the islands' earliest writers to leave an account of the kind of lives which, in Ó Criomhthain's prophetic words, 'will never be again'.

Their tools (the word 'technology' didn't gain popularity until the mid- twentieth century) spoke of a way of life that my own generation – which has sent people into outer space to explore the potential for life on such a hostile, lifeless planet as Mars – now considers absolutely impossible. For lighting they immersed a peeled rush in a scallop shell of fish oil. They used a simple, wooden tool for twisting straw into rope; a long-bladed spade with a wing on the side, called a sleán, designed specifically for cutting turf; and a wooden rake to help gather seaweed. To carry both the turf and the seaweed they made creels – wicker baskets that act as panniers – for their donkeys' backs.

From the Blasket Centre we walk to the ferry through Dunquin. It's a fine morning, but the only other people we encounter are inside cars, which are making their way to the bigger towns where bigger money is to be made. The school holidays haven't started yet, so Kirsty and I are two of the only tourists. Descending the steep path to the pier – up which the Islanders would have driven sheep, carried the corpses of their friends and relations, or hauled lobsters and mackerel – I can't help but absorb the sight before me: the Great Blasket

Island, looking like a monstrous whale basking in the Atlantic Ocean, its hump touching a clear blue sky as it faces the warmth of a sun awakening over to the east. From this vantage point it looks dignified, arresting, defiant, self-assured, entirely alone. Whoever lived there over the millennia, they knew how to look after themselves.

On the pier we board a small, lightweight ferry manned by two boatmen, who tell us they are fishermen on their days off. As we move across the Sound it rocks dramatically on relatively calm waters, and I can't help but contemplate what being in a canvas-covered naomhóg, *around midnight in the heart of a storm, must have felt like for a people who famously could not swim.*

~

The dawn sky is ablaze with all sorts of reds and oranges and pinks, so I pull on my boots and make off on a ramble while the weather is still fine. Red sky in the morning, shepherd's warning, and all that. As I walk I'm usually searching for something: berries, leaves, clarity, or lessons from the beings we are forgetting how to listen to.

The ungodly hours are my most sacred time of the day. I sometimes wonder why most of humanity refuses these moments, and places like this, but part of me is thankful that they do. The other part of me wishes everyone could witness the earthly glory before me, marvel at its mystery and worship it and all it comprises. Each to their own.

As I turn the bend by Packie's old house, which was once the gatehouse for Sir Thomas Burke's (and his predecessors') estate, I notice a few lengths of silage wrap wrapped around giant spears of Japanese knotweed, an invasive species which was brought to Ireland as an exotic ornamental perennial in the mid-1800s. There's more of the plastic on a hawthorn tree, where it clings to its own sense of meaning, while a loose ball of it is blocking up a ditch a little further up the road. In other parts of the country farmers have started using pink silage wrap in support of a cancer charity, but in these parts it is still just the black, run-of-the-mill, cancerous stuff.

A farmer across the way calls me over. He quickly needs a hand moving a couple of bullocks, and though it's only 1.5 kilometres up the road, he insists on giving me a lift. Ten years ago, when I was an environmentalist and animal rights activist, I couldn't have imagined a future for myself where I would be chasing bullocks into a pen, ready to be tagged and numbered and tested for tuberculosis. Back then, my opinions were derived from documentaries and footage from factory farms. These days I have broken sleep for two or three nights as I listen, first-hand, to the cries of the mother for the newborn calf that was taken from her.

When I lived in Bristol, activists – including me – would be forever falling out over theories of society, ecology, politics and culture. Out here we need each other too much to fall out over such things. The job takes longer than expected – the bullocks have their own ideas about the day ahead – and the farmer tells me a bit about his history, and that of the farm, as we go. It's a tough, unenviable story, and the more I listen to it the more I admire him for getting up in the morning and putting one foot in front of the other. It's the first time we've properly spoken to each other, and he leaves saying he'll pop over one evening for a drink in the *sibín*.

~

Over a year before I finally stopped using electricity altogether, I took my first baby step and quit social media. Like all good decisions, it was made in the pub. I had never actually liked anything about social media – the Silicon Valley companies behind it, the privacy and surveillance issues surrounding it, its ecological impact and antisocial nature – and yet I still continued to use it for work. Publishers like writers with a strong social media presence, I was told, and it was how most writers promoted their books, events and all the terribly important things they had to say.

It was after the third pint that I decided enough was enough. To say that it would have been too hard a decision to make when fully sober sounds overly dramatic – it's just a bunch of websites, after all – yet it felt a little like a carpenter deciding to give up his power tools. Whether I liked it or not, I knew my livelihood was aided and abetted by a handful of shadowy, too-cool-for-school corporations. I had often thought about quitting before, but my rational mind created all sorts of arguments for why it was a necessary evil. The stout offered a temporary shortcut to the soul.

The soul was much clearer on the matter. The soul said fuck it. The soul said stop buying in and stop selling out. The soul reminded me that I wanted to live entirely off the land anyway, and that any financial income would only hold me back and stunt my progress. The soul told me to live as I believed, first and foremost, and to let fate take care of the rest. The soul said enough. And then the soul, emboldened and thirsty, ordered another couple of pints.

I woke up the next day (with an unforgiving hangover), logged on to each of my social media accounts, and told people I was leaving. No big deal, I just wanted to say goodbye to the many people I would likely never see or hear from again, with a short explanation why. Lots of friends replied, imploring me not to quit, but instead to speak out about my views on industrial technology using the master's own tools. But I felt that there is only so long you can be critical of the very thing you are using, and that the best way of denouncing something might ultimately be to renounce it. Others agreed, and said they were thinking of signing off for good, too. I found the whole hullabaloo around someone leaving a few websites insightful, until I remembered how much I had struggled with the decision myself.

The following day I logged back on to each of my accounts and quit – it only took half the morning to figure out how to do it – and then went for a walk in the woods.

~

While I'm out working with Tommy Quinn, we get chatting about a session, a few nights previous, in a local pub called The Hill. It gets its name from the plain fact that it sits on top of a hill. The conversation moves on to the state of rural Ireland, and rural everywhere for that matter. He's lived here in Knockmoyle for all of his life, so his opinions on the subject hold weight with me. He asks me what technology I think had the most dramatic impact on life here when he was growing up. I state what I feel are obvious: the television, the motor car and computers. Or electricity in general. Tommy smiles. The flask, he says.

I ask him to explain. When he was growing up in the 1960s, he and his family would go to the bog, along with most of the other families of the parish, to cut turf for fuel for the following winter. They would all help each other out in any way they could, even if they didn't always fully get on. Cutting turf in the old ways, using a *sleán*, is hard but convivial work, so each day one family would make a campfire to boil the kettle on.

But the campfire had a more significant role than just hydrating the workers. As well as keeping the midges away, it was a focal point that brought folk together during important seasonal events. During the day people would have the craic around it as the tea brewed, and in the evenings food would be cooked on it. By nightfall, with the day's work behind them, the campfire became the place where music, song and dance would spontaneously happen. Before the night was out, one of the old boys would hide one of the young lads' wheelbarrows, providing no end of banter the following morning.

Then one day, out of nowhere, the now commonplace Thermos flask arrived in Knockmoyle. Very handy, Tommy says, and everyone wanted one. Within a short space of time families began boiling up their hot water on the range in their homes, before taking it with them to the bog. After millennia of honest service, the campfire was now obsolete.

It probably saved a fair bit of time, I say to Tommy half-heartedly. Aye, Tommy replies, no one had to go looking behind bushes for their wheelbarrow first thing in the morning.

~

Kirsty is using natural horsemanship techniques to train up a couple of horses to pull a gypsy wagon, as being gypsy-spirited herself she wants to spend time on the road in the way that best tells the story she wants to tell with her life. We were hoping to have the horses on fresh grass again by St Patrick's Day, but the weather decides to get wet again. Put them into a new field now and the poor-draining land around here will turn into a quagmire. They say Ireland would be a great country if only you could put a roof on it. The horses look longingly at neighbouring green fields, but patience is most critical at precisely the moment it is most difficult.

On my way up to give them hay, one of the neighbours reminds me not to forget to put the clocks forward tonight, remarking – as is required by convention – how it will be great to see a stretch in the evenings.

It has been a week since I've last known what clock-time it is. I know it's Saturday – well done, Mark, you're thinking – but that's about it. The clocks going forward tonight will make little or no difference to my life, most days at least. Tomorrow will be just the same as any other between winter solstice and summer solstice, simply a few minutes longer.

But tomorrow will be spring, and that's a whole other matter.

Spring

> I went to the woods because I wished to live deliberately, to front only the essential facts of life, and see if I could not learn what it had to teach, and not, when I came to die, discover that I had not lived.
>
> Henry David Thoreau, *Walden: Or, Life in the Woods* (1854)

Spring has sprung, and the song thrush and goldfinch are curating an exhibition of life, one which they've been working hard and thanklessly on, behind the scenes, over the long tough winter.

There's a certain magic in the air, and every living thing knows it. The old Irish had a word, *tenalach*, to describe the sense of connection you can feel with the rest of life on this annual day. Metamorphosis has the peacock butterfly foraging in my vegetable garden, frogs have abandoned all caution and are abroad in daylight, and the lambs in the next field up are leaping around in fresh pasture, lightening the hearts of hard men. I've a lot of work on – normal for this time of year – but I remind myself why I chose to live this way, and decide that nothing is too important to keep me from fishing this afternoon.

I call in to Paul, to see if I can tempt him into an impromptu ramble to a river. It's short notice – none – but the day is too inviting to refuse. There's a small river, called the Cappagh, across a swathe of fields at the end of his long *bóithrín*, and so we set our minds on exploring a hidden, disregarded stretch of it.

We settle by a pool where the Cappagh meets a tributary, the confluence of which appears to be the nexus for a complex web of life. A streak of blue-orange lightning we call a kingfisher darts downstream, while a heron holds court on the bank, ensuring that everything is as it should be. A monstrous, marvellous brown trout rises clean out of the water where a cluster of swirling flies roll the eternal dice of life and death. We've all communed, at this secretive spot, for the same reason. Food.

Unlike me, Paul has some experience with proper fly fishing, so while he is carefully selecting the right fly to use, I cast in with a small spinner, looking for something lurking in the banks. Fly fishing is an art; in fact it's more than an art. To do it well you need to become something of an ecologist. To know how to fly fish is to know your place.

All is calm and serene until *bam!* and something takes the bait. Through the clear water, I can immediately see the ferocious, Jurassic head of a pike. He looks sorely disappointed, as he thought he was onto a meal himself. Well, it is April Fools' Day, but I take no pleasure in cruel pranks. Nevertheless, I have an important role as a predator. Pike, ecologically speaking, are a problem in some lakes and rivers around Ireland. But because touring anglers prize them – for their size and photogenic qualities instead of their dead flesh and bones – the pike has inadvertently made fishing big business in Ireland.

As a thank you, the state has granted pike a protection of sorts, despite the fact that such protections are having an impact on other species. There are strict rules about the size of the pike you can kill, and within that range you are only allowed one. So while bottom-trawlers legally make deserts out of oceans, kill a pike over 50 centimetres long and you could be facing jail time. That is, unless you're the government, and the pike are in places where trout angling is considered to be more lucrative. Then it spends taxpayers' money to kill them. According to the ecologist Pádraic Fogarty, between 2010 and 2014 Inland Fisheries Ireland 'spent €725,037 removing 35,738 pike using a combination of gill netting and electro-fishing'. That's about €20 per pike. Hard to make it up, really.

The pike on the end of my line is slightly longer and heavier than the legal limit. I have a choice: to knock him over the head and break the law, or to go and buy a packet of bottom-trawled fish fingers sold legally from the supermarket for €3.99.

It's a tough choice. It's a time of tough choices.

~

Looking up into the abyss of what John Muir called 'cloudland', I notice the swallows are back, uttering their sharp 'vit' call as they go. A small congregation of males are twittering away on telephone wires connected to Packie's house, taking their rightful place in God's choir. They could be recounting tales of their winter adventures, or alerting the females to the presence of a magpie perched high on a distant tree. Who knows?

After a long sojourn in some particular part of Africa that I will never know, a couple of last year's migrants have travelled the thousands of kilometres back to Knockmoyle (53°05'N, 08°30'W) by way of their wits and wings alone. No satnav, no engines, no paper maps. They've nested in the same rafter of my woodshed as last year, and are getting themselves reacquainted with the place and prepared for the next turning of life's wheel. Though I have an inkling, I will never know why they've come back here, but it's a reassuring, life-affirming sight nonetheless.

We think we're intelligent, and at our best we can be. But from where I stand now, axing beech and birch in a woodshed, our intelligence looks no lesser or greater than the swallow's. No, it only looks different.

~

I hadn't seen an advert for the best part of a month, until this morning. An energy drink. It was on one of those A-shaped trailers that vans haul around cities all day. It came past our small-holding, I can only assume, on its way from one city to the next. The fifty- to eighty-year-old farmer demographic probably isn't their target market.

I studied marketing for four years, as part of an undergraduate degree in business. At the time I remember reading how each of us is exposed, on average, to around three thousand adverts every day – in shops, magazines and newspapers, on billboards, vans, radio and television. That was between 1996 and 2002, before the internet copy-and-pasted itself into every nook and cranny of our lives. I can only imagine what the figure might be in the digital age.

Seeing that advert felt strange, like a jolt. I can hardly complain, considering that my current exposure to advertising is tiny in comparison. Yet, as it drove past, its abruptness and overconfidence contrasted starkly with the woods behind it. If it didn't need permission to expose my mind to irresponsible, sexed-up market-ing for an unhealthy, addictive product, does that mean I don't need permission to put an axe through the advert the next time it drives by? My mind is private property too – perhaps the most private of property.

Home from the woods, I find a bank statement waiting for me in our letterbox. The bank is legally required to post one out every few months, whether I like it or not. Scanning through it, I notice that there is almost nothing in the debit column, which is good news, as there is almost nothing in the credit column either. The only work I receive any money for is writing, and everyone is telling me that quitting social media, phones and email isn't exactly going to boost my prospects with that. Everything else I do gratis, and always for something or someone I care about. Therefore the statement is mostly a small collection of bank charges which everyone in Ireland with less than €2,500 in their account has to pay – above that and banking is free. It's effectively a tax on the poor.

I put the statement on the fire grate and set some tinder and kindling on top, preparing it for the evening – better to do it now than when it is cold and dark – and take my breakfast outside to capture the morning sun. For possibly the first time in my life I realise that I feel content, without the desire for anything other than what's in front of me in that most elusive of moments, the here and now. I've been happy, hopeful and full of excitement plenty of times, but I can't recall a time when I was simply content.

~

The potting shed looks like Oliver Mellors has been in there with Lady Chatterley. There are empty compost bags strewn over tables, a few of those cheap, green watering cans dot the floor, and the plastic pots and cell trays which you use to germinate seeds are scattered everywhere. It's the start of April, so I need to get this place in shape for the next wave of the growing season.

Our plan is to grow enough vegetables to feed eight people over the coming year, so I begin the sort-out by making tables out of old wood and discarded pallets. Having already tidied up, I lay out a hundred black plastic seed trays across the tables, each containing twelve cells. All going well, that should allot us somewhere between a thousand and twelve hundred plants. I pack each cell with potting compost, give them a good watering, and plant one seed in each. I've planted everything from peas and pinto beans to all sorts of varieties of kale. You can't grow too much kale. On little plastic labels and strips of cardboard I write things like sweetcorn, Yokohama squash, echinacea, spinach, rocket, courgette, cucumber, swede, calendula, Brussels sprouts and beetroot. Before the day is out I've planted over thirty varieties of herb, vegetable and salad. I'm not a man to get too easily excited, but I'm already looking forward to the sight of the first seedlings bursting their way above the soil and into life. Such work is both satisfying and reassuring.

But then I make the mistake of thinking, and suddenly very little of it makes sense. Between the labels and watering cans and trays and compost bags it is wall-to-wall plastic. Literally. The walls themselves are plastic, the last of the polytunnels we inherited to remain standing. I know that food growing hasn't always been like this, but agriculture was the precursor to industrialism, and it was only a matter of time before they would marry. I've no idea what the word 'sustainable' means anymore – does anyone? – but this cannot be it. Nor do I have any desire to help sustain a culture dependent on plastic.

I know I need to become a much more proficient forager. Gathering plants in hedgerows, meadows and wherever they grow self-willed and naturally, without recourse to plastic or shed loads of tools – now that makes sense to me. I know I need to make leaf mould instead of buying potting compost, to focus on perennials instead of annuals, and a hundred other things to boot. But then I remind myself that these things take time – a forest garden takes at least fifteen years to get going – and that Rome wasn't demolished in a day.

Before I leave I water the plants, and marvel at how the wooden box of embryonic, fragile seeds in my hands will continue to keep us in vegetables over the coming year.

~

My fingers are tingling as I write. I've been out picking nettles to make soup for lunch, and dried tea for whenever. They're easier to gather than most people imagine. Such was their place in the diets of old that there's an old Irish rhyme which reminded children how to pick nettles:

> If you gently grasp a nettle,
> It will sting you for your pains.
> Grasp it tightly like a rod of metal,
> And it soft as silk remains.

If your mind is entirely focused on the nettles, you can pick them all day without getting stung. The moment you start daydreaming, they will get you good. Each leaf is a Buddha.

Secretly, I quite enjoy the sensation, and I'm told it's good for your circulation. The tingling mysteriously resurfaces at night, just as I take off my clothes for bed, reminding me a little of my day.

~

Up until recently, I had never really been one for walking just for walking's sake. Growing up in an industrial culture like mine, walking time was considered to be idle time, and idleness was not a virtue. It was something older people did, in an attempt to regain their health in retirement, and not something for young men who had careers to pursue, families to feed, businesses to grow and good health to deteriorate. That was before I read Thoreau's paean to 'Walking'. It's hard to ignore an essay that begins:

> I wish to speak a word for Nature, for absolute freedom
> and wildness, as contrasted with a freedom and culture
> merely civil.

After this, the art of walking became not so much a political act against the dogma and tyranny of ideas like efficiency and productivity, but a tradition worth keeping alive. As my legs began moving without any pre-planned, pre-conceived notions, I discovered something surprising, something at odds with what I had once imagined.

It's a wet morning – I mean Irish wet – so I decide to stay indoors and do some writing. But by lunch I'm restless. I remind myself that I am an animal, not a disembodied thinker, and so I follow the urge to go sauntering. I call in for Bulmers. He's never been trained, and isn't the sharpest tool in the box, so he needs to stay on the lead while he is on the road.

Actually, I'm being unfair to him. Instead of calling him untrained, I could have called him wild, self-willed, undomesticated, unbroken, all the things I would like to be again. He is more than capable of going off to the woods and the pasturelands for weeks on end, with nothing but the hair on his back, and making a good living for himself; which, let's face it, is more than I've yet achieved. But on the lead he's a pain in the arse, always pulling, chomping at the bit to follow his own heart's desires. It's exactly this indomitable spirit, this unrelenting enthusiasm for life, that rather annoyingly means I can never pass him on my way out for a walk.

We set off through what was, only a few weeks ago, a wood but which is now . . . well, I'm not sure what to call it anymore. The machines have finished and taken what they need to keep the machine economy running for a little while longer. At first I find my head still in writing mode, struggling to just be on a walk, thinking too much for my own good. That said, it reaps rewards of its own, as the fresh air and movement leads to breakthroughs on thoughts, sentences and paragraphs on which I'd got stuck. On which I always get stuck. I wasn't born a writer.

It's a kilometre and a half before I feel grounded and present. I notice deer tracks going into a few acres of adjoining woods. Noted. There are dandelions and nettles everywhere, both of which make good beer. Noted. The machines have left a trail of imperfect, unprofitable wood in their wake, which – in theory – looks perfectly profitable for my woodpile. Noted.

I finally make it back to the cabin, just in time, as the skies unburden themselves again. Enlivened, alert, clear, I work long into the night.

~

There's a megasylvan pile of old floorboards, rafters and joists in Mick's yard up the road. It's mostly pine and oak which has been ripped out of a house Mick's son is renovating, and he tells me to take as much as I like before it's made into a bonfire. It's low-hanging fruit compared to hauling logs out of the woods, and as this is the time of year to be getting wood in and dried, I take him up on the offer.

Rummaging through it, I sort out the good from the bad. Anything strong and without rot or woodworm I put into one pile, which I'll keep for any small DIY projects that may emerge over the coming year. The rest – mostly floorboards that were broken up as they were being taken out – goes into a woodpile. Many are still almost 4 metres in length, so I saw them in half to fit them sideways down the *bóithrín* to my lean-to, which is a good 300 metres away.

I knock on Packie's door to ask him if I can borrow his wheelbarrow. What are you even asking for? His is no ordinary wheelbarrow. It was handmade when he was a young man, and is still working perfectly fine today. It was designed and built with turf in mind, in the days before heavy machinery made such human-scale tools obsolete. The thing is imposing and without sides, meaning you would have been able to stack and move a lot of turf in it; as long, that is, as you had the shoulders to move it through wet, boggy land. Instead of taking ten to fifteen trips up and down the *bóithrín*, it will only take three with this contraption. Such logic can lead to quad bikes and tractors, though, so I remind myself that the law of diminishing returns applies to efficiency, too.

Back up at Mick's yard I stack the boards as tight and high as I think I can get away with. On each load I manage to get almost fifty 2-metre floorboards in, and the wheelbarrow is still not full. It's heavy though, and it takes the best part of me to make it safely to the lean-to without it tipping over. As I walk up the hill for the third load, I feel a renewed sense of admiration for the folk who would have once done this, from dawn until dusk, for days on end as they took turf from the bog.

I spend the rest of the morning on the sawhorse, where I break it up into stove-sized pieces. Anything thick is cut up with the bow saw, while the sledge-hammer takes care of the rest. It's all done before hunger gets the better of me, and I estimate that I have about six weeks' more wood stacked up.

I stand back to admire the woodpile. The work was not as satisfying as spending the day in the forest, and the stack itself reflects that. It's just a pile of smashed-up timber and not the chopped and well-stacked row of spruce, beech and ash that I could look at all day.

Still, wood's wood.

~

On the same day in 1996 that all of my boyhood mates took off for university in Belfast, I got on a bus destined for Galway City and, for the first time in our lives, our paths diverged. We had known each other since we were kids, and we had seen each other through the turmoil of adolescence. We played football together, and won and lost together. We got in trouble together, and got ourselves out of it together. They were my tribe.

There was a sadness about leaving. Even though we had planned to see each other at weekends, part of me knew that we would never again all live in the place which we had always considered home. Moving to Galway was like starting life over again – there was no history, no ties, no bonds, no familial connections and no reputation to rely on. Up until that point, I had only been to a city – what farmer and writer John Connell calls 'those farms of men' – a handful of times, and suddenly I was living in one. It was scary, and exciting, like many scary things are.

It soon turned out that I had about as much interest in being in lecture halls as I'd had in classrooms. I found myself taking subjects like economics, IT, accountancy, maths, retail marketing, labour relations and entrepreneurship, but my heart wasn't in any of it. The problem was, having now left my hometown, I no longer knew where my heart was.

It was because of this that I was found more often in pubs than at lectures. Or perhaps it was because I had suddenly come into a bit of money. When I was thirteen, one of my best mates and I were hit by a car travelling at 60mph. I was considered dead. Those who had come onto the scene had put a blanket over me, as I lay motionless against a concrete post, while they tended to my friend, who was semi-conscious and badly broken in the middle of the road. The woman who had hit us had been trying to overtake, and when she realised that she couldn't get back into her own lane before meeting with oncoming traffic, she cut across into the hard shoulder, where we were walking. Within three days, however, I was out of hospital with no broken bones (my friend was out of action for over a year), and I was eventually awarded £13,000 (Irish punt) in compensation. It's odd, I don't remember having had any sense of being awarded another shot at life.

That was an awful lot of money for a young man who hadn't had two pennies to rub together before. Yet on the day the money was transferred to my bank account, shortly after my eighteenth birthday, something in me didn't want anything to do with it. I don't know why. It wasn't like I had critiqued the concept of money at this point. I just didn't want it. I had planned to give it to Oxfam, but my mother convinced me to hold onto it until I was sure. She was only looking out for my best interests, as that had been the most money any of us had ever had. So I drank it instead, and made sure that my new mates in Galway drank it with me.

By the time I'd started second year I felt the whole thing was bullshit, a big waste of time, and on a couple of occasions had decided to pack it in. I remember once quitting and going for a job with a window cleaner. He told me at the end of the interview that he would be happy to give me the job if I wanted it, but implored me to finish my studies instead, and to make any decisions after that. There was something about the look in his eye that held weight with me that afternoon. So I went back. A week later I lost the floppy disc on which I had saved a three-thousand-word essay I was due to submit, which felt like the final straw, so I quit again. This time it lasted a few hours until, walking through a field I hadn't recalled being in before, what did I find lying in the grass, fully intact, but the floppy disc. I took the hint and decided to finish out the year, and to leave taking stock until afterwards.

Miraculously, I made it through second year, passing one exam on repeat. I was glad I had stuck it out, but I needed a change of scenery. I decided to take a gap year. What I would fill the gap with I wasn't sure. Up until then I had lived for football, and the feeling that you would live and die for your tribe, on or off the pitch. But as I fell away from football, that sense of togetherness slowly slipped away with it. I felt uncertain in the world, like I didn't quite know who I was or why I was here.

I wanted meaning and purpose back in my life. So what did I go and do: I got a job on the assembly line of an American pharmaceutical company on an industrial estate in Galway, and went out drinking every night with the wages.

~

I spend the morning hunkered down in the herb garden, weeding. I find it repetitive, quite boring work, yet I find it affords me plenty of time to think. I use that time to wonder what my hunter-gatherer ancestors would think of me now, as I spend my time picking the plants I *don't* want.

I accidentally slice a worm in two with my trowel, and it writhes around manically, in much the same way I would if someone speared me. It isn't the first and it won't be the last. Does that mean my herbs, or anyone's herbs for that matter, are still vegan, or is a worm's life of less value than a deer's? Some of the plants I pull out I know only too well, while others I don't know well enough. There's a certain arrogance to the job that I'm not entirely comfortable with. Without understanding their qualities, or their place within this place, I defame some plants and call them weeds, while others take their lofty place as herbs. To those whose qualities our culture understands, we grant life; the others we kill. But one man's weed is another man's medicine, and only some phenomenon greater than myself knows the real purpose of anything.

I know many excellent gardeners who frenetically pull out nettles – a weed packed full of nutrients, flavour and potential for high-protein leaf curd – while simultaneously struggling to keep slugs off their lettuce. Dandelion and horsetail suffer the same fate, all for the want of a little understanding.

Over lunch I sit and watch the little red face of a goldfinch bob up and down on a dandelion flower, and he (the male's face is a more vibrant red, the yellow streak on his wings a little wider) is soon joined by a mate. In close attendance is the honeybee, who is getting on with business in among a patch of primroses and blue-bells. Everything knows its place. I need to be careful I don't forget mine.

~

I'm out working in the garden when Jim, a local farmer-cum-road worker, stops in for a chat. He asks me if I know that it's Good Friday, a sacrosanct bank holiday in Ireland and one of only two days in the year that even publicans are forced to take off. I tell him that I didn't even know it was Friday, let alone a particularly good one. He thinks I'm joking, and laughs.

He tells me that this will be the last year the pubs will be forced to close on Good Friday. The Easter weekend is a big, international holiday break, and politicians and business leaders in Dublin had successfully argued that the capital was losing millions by not having pubs open on this religious holiday.

~

The boat ride from Dunquin to the Great Blasket takes around twenty minutes, with barely any human effort required other than the starting of the engine and the steering of the wheel. Being out of tourist season, we are alone, but I imagine it could ferry a hundred people without having any impact on the speed of the journey. The naomhóga *which the Islanders travelled to and fro on, using oars and carrying up to eight people, would take anywhere between forty-five minutes to an hour, depending on the conditions and the load, which ranged in weight from a boatful of mackerel to the heavier burden of a child's corpse.*

It is a calm day, and we have an easy landing. I remember reading archives of stories about Islanders getting caught out in merciless storms and being unable to land at all. We ascend the steep, grassy path from the rocks, and I make a beeline for Tomás Ó Criomhthain's house, or whatever is left of it. As it turns out, there is more than expected. The Office of Public Works, which generally does good work, is renovating it. Whether or not such ruins should be restored and preserved for eternity, or allowed to return to the earth, is a matter of opinion. My main reservation with this particular renovation, as I stand examining it, is that the builders are using cheap, mass-produced materials, giving Tomás's old abode the tinge of bland suburbia. They certainly aren't using the flotsam of shipwrecks, which the last human inhabitants often used in the maintenance of their homes. Instead, they're helicoptering in the materials. Looking at the skill in Tomás's own stonework, it strikes me that they are in danger of preserving the fact of the cottage while doing nothing to preserve its spirit.

I follow the tourist map around the other houses. You can still, just about, make out the web of pathways running from house to house. Robert Macfarlane and Roger Deakin have, in the past, drawn attention to how such tracks are signs of connection. Today, however, these paths are not kept worn and alive by economically interdependent villagers, but by day-trippers between whom there is no real connection beyond the passing moment. Following these tourist-trails, I find myself visiting the houses of its famous writers first, though all of the houses were made of the same stone. I walk to the childhood home of Muiris Ó Súilleabháin – author of Twenty Years A-Growing *(Fiche Bliain ag Fás) – next, and then to that of Peig Sayers. But it suddenly feels quite wrong to do so, like I am getting myself caught up in a strange variation on the cult of celebrity. So instead I visit the village well, the gathering house (An Dáil), the unconsecrated graveyard, the tiny school and post office, along with* Tigh na Rí, *the house of the last Blasket King, which appears to be the smallest house of all.*

It is at Peig Sayers' house that I meet a man called Diarmuid Lyng, who stands out among us tourists. He's wearing work clothes and has a rough wiry beard, while his face has an open, friendly look about it. Having walked past other visitors silently all day, I find myself saying hello, and we get to talking. He is out here volunteering, working on Peig's old house. I later find out from the boatmen that there was a TV documentary aired about him a few months earlier. Apparently he was a well-known hurling star, but – in the words of the ferrymen – he 'lost the plot', gave it all up and moved out to West Kerry, where he now spends time volunteering on the island. As we speak over the course of the afternoon, I see nothing in the contours of his face that suggests he has lost the plot at all.

It transpires that he is a good friend of a good friend, and that we have many other friends in common, something I'd oddly already suspected. He tells me that, contrary to what I had been led to believe, visitors are allowed to camp on the island; that is, if they are hardy enough to want to. We haven't brought a tent, thinking there to be no point, so he invites us to stay the night in one of the old houses that, despite being extremely basic, still has its roof and walls intact. It's more than good enough for us. We offer to help him with his work for the evening, and I feel excited about how the adventure is unfolding. But just as the last boat is taking day-trippers back to the mainland for the evening, he decides to check with his boss that our staying the night isn't a problem. We had all spoken at length earlier in the day, sharing many common interests and perspectives, so he had presumed that it wasn't going to be an issue. He was wrong.

The owner – a friendly, thoughtful man – tells him that our staying the night is an insurance risk, and that we therefore have to go back on the last boat that is still waiting down at the landing. Diarmuid looks frustrated, but I tell him we understand, and that this is just the way of the world at this brief moment in time. I laugh – not a joyful laugh, but a laugh nonetheless – and wonder if I am the first person to be refused a bed for the night, here on the Great Blasket, on insurance grounds.

With that, we run down the hill to the landing area, where the good-humoured ferrymen are patiently waiting on us. We get off the boat and start walking the long road back to Dingle. I have a feeling I'll be back again.

~

Cuck-oo, cuck-oo. Forget the Gregorian calendar, I hear the first cuckoo this morning, and that means that it's the third week in April. The male's song feels serene and calming – to me, that is. For songbirds, it is a grave warning. Soon one unfortunate little bird will return to her nest to find her own eggs gone and the cuckoo's egg in their place, and she will care for the imposter until it hatches.

There's an old superstition in Britain and Ireland which says that if you have money in your pocket when you first hear the cuckoo, you'll never be without it for the rest of the year. I'm wearing just shorts today, and I don't even have pockets.

I stand by the spring for a while, hoping to catch a glimpse of him, before eventually leaving disappointed.

~

There's an unusual letter in the post. It's from a stranger who writes that he has recently quit his job as a senior claims manager for an insurance company in Australia, and says he is now looking for work he would find more meaningful. He explains a little about the pressures and criticisms he has received from family and friends because of his decision, and wants to know if I have any advice. I'm a poor career guidance counsellor, but I tell him to follow his heart. He goes on to tell me about how his search has uncovered work in what is now being called 'the spiritual services' – menopause mentors, death doulas, walking companions and so on. Apparently demand is growing for such things.

You know that industrial capitalism is nearing the completion of its ultimate vision when people have to pay their neighbours to go for a walk with them.

~

With the exception of 'the long farm' – oldspeak for the grassy verges along the sides of *bóithríns* which serve as some of the last vestiges of commonage in Ireland – it's never easy to find grass for horses in April. It becomes clear to me that they need a paddock on our own smallholding, which means I need to get fencing, something I am disinclined to do. Electric fencing isn't an option, so I opt for post-and-rail on a shoestring instead.

For the rails, I want to use up some spruce boards I have left over from the cabin build. Having only eleven boards and needing twenty-two, there is good news and bad news. The good news is that the boards are twice as wide as they need to be, and therefore can be cut in half. The bad news is that these boards total 80 metres in length, and will have to be cut with a handsaw along their longitudinal axis, something wood is never keen on. That's an awful lot of sawing.

Needs must. I decide with Gillis – still here – to take one board at a time, each starting at opposite ends and aiming to meet in the middle. Starting early, we go at it as hard as we can, focusing on each board in its turn. Despite switching arms regularly – it's important for the physical balance of your body to learn to cut straight with both arms if you do a lot of sawing – I can feel them by the time we're finished in the afternoon.

I hang up the saws in the tool shed. On the floor I notice a bandsaw that Tommy left here a few years ago. Half-broken, it still would have done the job in fifteen minutes, with little effort on our part. I had thought about it a few times this morning, and I'd be dishonest if I said I hadn't been tempted by it, lying there, like a siren, luring me in with all its electrical glory. But I have made my bed, and feel content in my monogamous relationship with hand tools.

~

The machines are back, in a different part of the woods this time. I can hear them as I write. They don't stop, from dawn until dusk. There were paths I had made in those woods, my own secret pattern by which I could roam and explore. There was one little holly sapling, among the stand of spruce, which I had watched grow over the last two years and had for some peculiar reason become attached to. The landscape is so utterly changed that I wouldn't even know where to look for it now.

But I suppose it's jobs, prosperity and growth. What kind of jobs, prosperity and growth, I'm not sure. It's certainly not growth for that little holly.

~

May Day, and the orange-tipped butterfly, buff-tailed bumblebee and the German wasp – all proud, hard workers (or is it play to them? Or simply life?) – are on riotous form. I'm awoken by the sound of defiance, irrepressible joy and small talk, all chirping, twittering and singing their way through the open window above my head. Feeling too enlivened to go back to sleep, I decide to make the most of the sunrise and to go foraging for broadleaf plantain (*Plantago major*) for the first time this year.

Like many hayfever sufferers, for much of my life I've looked forward to June and July with mixed feelings. Hayfever plagued my childhood. Glorious sunny days were often spent sneezing indoors with a wet towel over watery, itchy eyes, along with a nose full of snot on the inside and a parched red look on the outside. It made making, and keeping, friends more difficult, especially in my self-conscious teenage years when it was at its worst.

It used to be that people who suffered from hayfever in their youth grew out of it later in life, but with increasing levels of carbon in the atmosphere promoting pollen production in trees, we are now seeing the trend in reverse: people who never had it in their youth are developing it in their thirties and forties.

I tried every product in the pharmacy when I was growing up. Nothing really worked and it felt like most of them just made me drowsy. By eighteen I'd had enough, and went to my doctor to get a steroid injection for relief of the symptoms. He told me that it would work for three years. He was right. Life improved until I was twenty-one, when it came back worse than ever. At that point I resigned myself to buying the tablets again, and basically learning to live with it.

At twenty-eight, when I began to live without money, I no longer had any way of buying anti-histamines, so all of a sudden I was forced to find a new way of dealing with it. That was when a visitor told me about the secretive qualities of broadleaf plantain, a robust and determined weed that you will often find growing up through the cracks of pavements. It's a natural anti-histamine, and it starts to emerge at the beginning of May, which is precisely the time you want to start taking it. She told me that staying unstressed throughout the summer months, and avoiding dusty or polluted environments, would also help my body respond better to what it considers to be an attack.

It worked. Not entirely, but within weeks it became more like a blocked nose in the morning, and the odd sneeze, rather than something which dominated my life.

Back outside, in the 'farmacy', the plantain doesn't appear to be widespread yet, but I come across a patch here and there, which is enough to get me started. The leaves are still young, so I pick only one leaf from each plant, otherwise I may stunt its ability to photosynthesise and flourish. Patience now will benefit us both in the long run.

I basket about twenty leaves, take them to the fire-hut, and boil up enough water on the rocket stove for a large teapot. While it brews – it needs two hours upwards, ideally – I make myself a cup of chocolate mint picked fresh from the herb garden, lie up against an old willow tree, and watch the world go by. I've a lot of things on my mind to do but, for medical reasons, I decide that it's best to just lie here for a while. The two wood pigeons in the Scots pine in front of me are doing much the same.

A robin comes up close. I recognise him from his breast markings and stout shape. He has eaten worms out of my hand before, and has a character not unlike Packie's: cheeky. He's hoping as always for food, but he's out of luck for now; like any ambitious boss, however, he's keen for me to get back to work soon. Up above a raptor – is it a hen harrier? Hard to say from here – is loitering with intent. For a creature down below, the world is going to end on this beautiful day.

~

It's my birthday today. Thirty-eight. I wonder if I'll have a mid-life crisis soon. Most people probably think I'm already having one.

I'm not one for making a thing about my birthday, or so I tell myself. But even so, I would usually get a feed of well wishes on social media from people who only knew it was my birthday because Facebook sent them an alert (though, strangely, it was still nice to have felt remembered). Family and friends would call or text, all asking me what my plans for the day were, or encouraging me to get drunk. The usual stuff. Mum would phone religiously. I'd reply with things like 'oh it's just another day' or 'I'm trying to forget them at this stage,' and other half-truths.

Today I hear from no one, so it's not hard to forget that it's just another day.

~

The orchard looks a mess. Disorderly. Wild, even. As every spare moment last year was taken up with building the cabin, it has been almost twelve months since the orchard has been mown. The grass has now matted together, while thick tufts of rush and last year's dead weed stalks draw the attentive eye away from the usual orchard aesthetics.

I grab my scythe. I probably should sort the situation out, I tell myself, and make the place photogenic again so that when visitors come here it looks something like what they are expecting. The scythe is sharp, and ready to be swung systematically up and down the field. Used well, it can be just as effective as a strimmer. In fact, in the annual scything championships, some of the contestants with the scythe can actually beat the contestant using the strimmer. Used badly, however, it can be a slow, dull pain in the arse. Keeping it keen-edged is key to pleasant, effective, skilful scything. As with many things, if you try taking shortcuts it will take you much longer.

I've barely begun when a well-fed frog leaps over the blade, and not a second too early. I bend down to see him better, and it suddenly dawns on me how many more frogs there were here when we first moved in and began acting like we owned the place. Beforehand it had been more or less abandoned by humanity for over five years. The whole thing reminds me of Chernobyl, and how the wildlife there has been faring better from living with the fallout of a nuclear disaster than it was with industrial man.

I drop my scythe, and go exploring instead. There are forget-me-nots in places I'd forgotten. Creeping buttercup – the agriculturalist's adversary – has taken over in patches, and in among it I find bees – the agriculturalist's friend, though the fact may be forgotten – enjoying a food source that hadn't been available to them a year earlier. I finger my way through a section of matted grass and discover a web of insects I hadn't, until now, paid the slightest bit of attention to. They seem to be getting on with the old, art of survival. A few metres from the hedgerow I notice that knee-height saplings have burst through the earth's skin in places where I had been too busy to scythe. In fact, these saplings are everywhere; there must be thirty, forty, fifty around the edges alone. I had planned to organise another round of tree-planting for this coming November, but now there is no need. The land knows what is best for itself much better than I ever will and, better still, it will do the job for free and with no effort from me. According to Peter Wohlleben, a German forester and author of *The Hidden Life of Trees*, this wilder approach gives the trees that do come up a much greater chance of becoming ancient, too.

Further up the field I notice docks, which I've always found unsightly. Modern farmers hate them, but they oxygenate the soil, something which does no harm at all on these heavy, compressed clay lands. Maybe the docks understand something we don't? I look at the rush. Again, it's ugly when viewed through eyes that have become accustomed to manicured gardens. Yet I know they give me candle wicks, and once provided thatch for roofs. I've no doubt they provide other ecological functions, too, I just don't understand what they are.

This land I call mine clearly wants to be woodland again. Maybe I should start listening to it better.

The apple trees (*Malus domestica*) I planted four years ago are back in leaf, and now doing well. They wouldn't complain if they got a bit more care – what domesticated creature would? – so I put the scythe away, grab the fork and wheelbarrow and make for the compost bay instead. Like all things domesticated by human beings, the grafted apple tree is dependent on us, which means that, as the old ways pass into extinction, it has become dependent on industrialism, the very thing that is now making the climate less hospitable.

~

It's hard to escape the news. Packie tells me there has been another terrorist attack (I wasn't aware of the others) in another European city. He couldn't recall which one. Someone drove a truck into a throng of people, killing many. The world's gone mad, he says. I nod. The world has certainly gone mad.

One part of me feels like I should keep up-to-date with important global affairs such as this. The other part feels like I would be better off calling in to see if Kathleen needs a hand with anything.

~

There's a real battle going on up above. A song thrush and a magpie tumble around in the air; the magpie – the original aggressor – is now doing its best to escape, while the song thrush – who was minding its own business – appears hell-bent on ensuring that the message is clear: fuck off, and leave my nest alone. What drama. No police, no courtroom, no victim support. Just life, lived immediately and directly.

The advantage swings. The magpie makes a different call, to which its mate responds by flying in the direction of the nest, near which the battle first begun. There's an old nursery rhyme about the magpie, which starts 'One for sorrow, Two for joy', but it doesn't hold true for the song thrush. Outnumbered and outflanked, it races off in the same direction and, despite its physical disadvantage, somehow succeeds in chasing the second magpie off too. The nest is safe. For now.

Down below, another song thrush has arrived and, taking three small jumps at a time, scours the soil for worms, insects and other living things that are minding their own business.

In between sky and earth there is man, chasing success, scouring his new asphalted terrain for meaning, driving to the supermarket for milk and cereals, bacon and diesel, convenience and value-for-money, elevated above the savage violence of barbaric creatures like the song thrush and magpie.

~

I hand the postmaster my letters for the week, and he tells me that the price of stamps has just gone up by almost thirty per cent. I do some quick maths. The cost of the twenty or so letters I currently post each month is equivalent to my old mobile phone bill, and roughly half the cost of an internet connection, the prices of which are going in the opposite direction to the stamp. Rumours persist that the postal service is struggling financially, but no one seems to know yet what the upshot of that will be.

On my way out the door, I notice a couple of posters on the noticeboard: one is a fundraiser for a local table tennis Paralympian, another for a trad session in a pub up the road. Must go to those. I overhear a couple of men talking about Brexit and, despite being two farmers living many hundreds of kilometres west of Britain on the Atlantic coast of Ireland, the impact it might have on their livelihoods. They disagree. One says he'd have stayed in the EU, and that he thinks there's a lot of racism going on, while the other says he 'couldn't give a shite' if it means a few quid less or not, and that he would have pulled out too, given the chance. He's sick of being ruled by Merkel and bureaucrats who know nothing about life on a small farm in rural Ireland. On that point they do agree, and laugh, and arrange the loan of a trailer later.

~

Eugene offers me a whiskey. The half-light of dusk is fading fast, and I should be making inroads into the 20-kilometre cycle home, but I already know that resistance is futile. Eugene's a farmer, and he lives next to one of the lakes where I fish. One evening I met him on his bike – a normal push bike fitted with a taxless 49cc engine – and we spent some time talking as he rounded up his cattle. I told him that the next evening I managed to catch a few fish I would drop him in one.

He's surprised to see me again. Fine fish, he says. Not quite as fine as the one in my pannier, I tell him. As he pours me and a friend a couple of large glasses – doubles? Triples? – he tells me that having 'the word' (shorthand, in these parts, for doing what you say you'll do) has become an endangered trait, and that he hadn't expected me to actually drop one in.

We get talking. Shaking his head, he tells us that he would love to go fishing in the evenings himself, but that he works every spare minute God sends. Most of his income goes towards paying off loans he took out for all the farm machinery he invested in. Without the machinery, he says, he can no longer compete, such is the nature of farming today. His grandfather, also a farmer, was a keen fisherman, however, and loved nothing more than to go out on Lough Derg.

We have a parting glass, and he walks us down his drive. We pass his new digger, and I think about the spade I bought for €5 at a car boot sale. It has a wooden handle, is at least twenty years old, and has a fine edge on it.

~

It was 1999, and I had just spent seven months working on the assembly line of a pharmaceutical factory. The repetitious nature of the job gave me much time to think, yet I was no closer to figuring out what I wanted to do with my life. My old friends were all planning to go out to New York for the summer, to work and play football, so for the want of a better idea I decided to do the same.

The first thing you notice about New York is that everything is for sale. The place is the economic theory of division of labour taken to its logical conclusion. As a labourer renovating pubs at night, I usually worked six days a week – often seven – and sometimes found myself doing twenty-four hours in one shift. Being illegal and skint, I had no other option. Not that being a US citizen necessarily makes life any easier in New York. Many citizens only get two weeks' holiday per year. It's a money-rich, time-poor city, the upshot of which is that most people do their week's work and then pay for everything else with their earnings. I mean everything.

I can't remember once cooking in all of my time there. I was usually too tired. When I ordered pizza over the phone I could ask the delivery guy to pick me up toilet roll or toothpaste on his way. As long as I tipped well enough, it was never a problem. If necessity dictated that I use my own legs, by the time I'd made it to the closest grocery store I could have bought drugs, souvenirs, cleaning services, sex, a manicure and about ten different types of fast food. The shop was only 50 metres away from our flat.

Things in New York went badly, quickly. My first job was the worst I've ever had. It was an Italian crew, and they hated the Irish, who had historically competed with them for the same shit jobs. We still do. Sometimes they would throw things – chairs, tools, bars – at me and the other 'micks', allow us virtually no breaks, and regularly warned us that they were going to pop a few caps in our asses. It was like being in a poorly produced Mafia movie. It wasn't pleasant, but I personally didn't find their attitudes towards us racist. It was simply economic.

To add insult to injury, as I went to get off a busy train after my first week's work, I realised that my bag had been stolen, seemingly from right under my legs as I dozed. I felt like a fool. Everything I owned was in it – all of my clothes, my first week's pay, my passport, my camera, and the number of a girl I had met the night before. All I had left were the work clothes on my back. The friends I had come over with loaned me some clothes and cash, and we went to an Irish bar in the Bronx where your accent was still as good as any ID.

I hated New York, and I was beginning to hate cities. I had no idea what I was even doing there. I wanted to do something that had meaning to it. But what? And what did meaning even mean?

After a summer in New York I left as skint as I had arrived. I decided to go home, finish my degree, get my head down, and take it from there. What I did know was that New York's money-rich, time-poor way of life was not for me, and for that lesson alone I was grateful.

~

I miss Liam Clancy. I miss Joni Mitchell and Luke Kelly and Pearl Jam. I miss hearing Ewan MacColl singing about 'The Joy of Living'.

There's something strange and unbalanced about missing people who you've never even seen alive and in the flesh, people who don't know you exist, or care for that matter. I 'got to know' these people through audio and video recordings, but know nothing of what they are – or were – like as people. Many of the musicians I grew up listening to are dead, preserved for posterity by the magic of electricity, through which people continue to enjoy them as if they had never died. The moment I decided to reject the immortalising world of television, radio and the World Wide Web, their voices and music suddenly followed them to the grave, as far as I was concerned. It was if they all died on the same day. It can be a sad thought, whenever I think about it.

Saturday night and The Hill is packed. There are ten local musicians playing, all from around the village of Kylebrack, complemented and supplemented by those in the crowd who, throughout the night, like to offer up a song to a silent, attentive audience, as is the tradition in these parts. The quality of the music, and the singing from the audience, is good, though nothing like that of the greats. I sometimes wonder, however, if our constant exposure to the world's best has degraded our relationship to our ordinary, local musicians in the same way that exposure to porn stars with fake 34DD breasts and 10 inch penises has degraded and damaged our sexual relationships with those ordinary men and women we call our lovers, our husbands and wives, our girlfriends and boyfriends.

Michael, from down the road, is on the accordion. Last week Kirsty and I spent the morning catching two horses that had broken out of a field which was clean out of grass, and the rest of the day trying to figure out who owned them. It turned out they were his. He offers us a few pints for the bother, but we tell him that it was none, and that most of Knockmoyle would be drunk if we bought drinks for every kind act that had been done for us since we moved here.

The owner of the pub, and a few of the locals, ask Kirsty if she'll do some hula-hooping for the next couple of songs. She'd love to, she says, and before the night is out most people are up dancing too. One of the old boys has a short-lived go with the hula-hoop, for which he gets the cheer of the night.

~

They say that if you feel you don't have fifteen minutes to meditate each day, then you need to do an hour. I'm sure it would do me no harm at all, but I've never been much of a man for sitting cross-legged, focusing on my breath. Instead, I prefer to whittle.

Whittling is a form of practical meditation, which pre-dates the Buddhist and Hindu civilisations. It's as simple as it gets. To make a tablespoon you take a branch – I prefer green birch, but holly, beech, maple and cherry can work well. Avoid softwoods. Saw it to length, axe it in half, draw out the shape of the spoon you're aiming for, and start whittling it away with a small carving knife.

Your knife, along with your sense of awareness, needs to be sharp. Drift away in your thoughts, worries or daydreams for one moment and, if you're lucky, you'll shave off a sliver of wood that may take you twenty minutes to correct; in the final stages you may not be able to correct it at all. If you're unlucky, you may shave off a sliver of flesh from your finger that may take a week or two to correct itself. Nothing focuses the mind better than blood, or the thought of showing the woman you love an ugly, impractical spoon.

Sitting by the rocket stove in the fire-hut, tending to a brew, I put the finishing touches to a soup spoon. It's not perfect, yet every imperfection tells a story of my afternoon, which makes it perfect to me, and me only. When I eat soup from this day forth, that small dent in the bottom will be my Buddha, but I'm content with it. There's no point being otherwise.

~

When you consider that, in the mid-1900s, the Great Blasket Island became its own literary sub-genre – albeit a now largely forgotten one – it is surprising how little we know about the island's history. Even its name is a bit of a mystery. The receptionist at the Blasket visitor centre felt the island was long overdue an archaeological dig, to find out more about its ancient past. I'm not convinced that the wildlife that continue to call the island home would agree with such civilised opinions. Personally, I'd prefer to understand the life that lives there now than that which may have lived there thousands of years ago. It's all too easy to destroy the present while exploring the past or the future. I left the island hoping that a dig would be one of those big expensive jobs that the government, or its private partners, would never get around to doing. In a world of disappearing wildness and wilderness, some things are better left mysterious.

One of my immediate impressions of the place was that, despite being 5 kilometres off the south-west coast of Kerry, the island was once connected physically to the mainland parish of Dunquin, and not just socially connected as it was between the early 1800s and 1953, when the last remaining Islanders were eventually evacuated. I would later learn that geologists had confirmed what their own eyes would have suspected. This compulsion 'to know' places which we show no interest in actually getting to know reminds me of E.O. Wilson's words:

> If there is a danger in the human trajectory it is not so much in the survival of our own species as in the fulfilment of the ultimate irony of organic evolution: that in the instant of achieving self-understanding through the mind of man, life has doomed its most beautiful creations.

Whoever the first inhabitants were, they left behind stone beehive-shaped huts called clocháns, which would later be used by monks. It was probably these monks, or the Vikings who followed, who built the promontory fort at An Dún, one of the island's peaks.

I've since wondered whether life would have been harder, or easier, for these earlier inhabitants than it was for those who moved there at the start of the Industrial Revolution. On the one hand, they wouldn't have had the advantage of the seine boats which transformed fishing around 1800; on the other hand, these ancient frontier people wouldn't have had to compete with the titanic British, French and Spanish trawlers that would quickly and efficiently empty the Blasket Sound of its mackerel, and other fish, to feed their own distant, growing populations. As later generations would come to learn more acutely, in a globalising world, technology offers an advantage only for as long as you are the people with the best technology.

These monks and Vikings would have had the added advantage of not having to give a portion of their lives to paying rents to landlords who, by the eighteenth century, had mostly inherited the land from plundering conquerors. It was because of such land rents, which were hiked up all over Ireland around 1800, that the Islanders fled mainland Ireland to take refuge on this remote island in the first place.

Nowhere, it turned out, was safe from these extortionate rents. The first generation of Islanders in the early nineteenth century were charged a rent of five pounds a cow, the severity of which is emphasised by the fact that their grand-children would only be asked to pay a fifth of that, and even at that they could barely afford it. Regardless of the amount, what each generation seemed to have in common was a determination to resist paying the rent to the Earl of Cork. Their efforts were notorious.

On one occasion a ship of bailiffs, armed with guns, anchored in the bay and attempted to land on the island and take whatever they could – fish, cattle, whatever money or belongings the Islanders had. According to Tomás Ó Criomhthain in The Islandman, *many of the Islanders feared that there wouldn't be 'many houses left in this island by evening'. That was until the island's women gathered their own ammunition – rocks – on the cliff above the landing spot, and waited for the first bailiff to set foot on their shores. 'Won't you be better off dead than lying in a ditch, thrown out of your own cabin?' one* woman asked. As the bailiffs attempted to come ashore, wearing dark uniforms and high caps, the women – each of whom had a gun fixed on her – showered them with rocks. One woman was so incensed and convinced they were done for, that she had to be restrained from throwing her baby at them as soon as they were out of rocks. The bailiffs, stunned, retreated back to their ship before making two more attempts to land. But the women – rearmed with stones gathered by the men – held their ranks, and not one stirred. Ó Criomhthain later said that 'they felt less fear than they inspired', and so 'the ship cleared off with all of its crew that day without taking a copper penny with them'.

On another occasion the Dingle police confiscated the Islanders' fishing boats while they were in town selling their wool, sheep and pigs. Acting on behalf of the Earl of Cork, they demanded the rent for the whole island before they would return their boats, which the fishermen depended upon for their livelihoods. Friends from Dunquin and the surrounding area offered to contribute money towards this demand, but the Islanders refused with thanks. Instead they decided to leave the boats with the rent collectors and go back to fishing with their naomhóga. The bailiffs later tried to sell the boats, but as not one person would buy them they eventually rotted in a field. 'The upshot of the whole affair,' Ó Criomhthain said, 'was that you wouldn't find it hard to reckon up all the rent we've paid from that day.'

It was out of this tough, uncompromising crop of Islanders that the Blasket literature emerged. To call it an unlikely genre would be understated. Peig Sayers – whose autobiography Peig was required reading on the Irish school curriculum until recently – could neither read nor write; she dictated it to her son, the poet Micheál Ó Guithín. Hers, after all, was an oral culture, while she was one of its most gifted storytellers.

If it hadn't been for two Englishmen, Robin Flower and George Thomson, who regularly visited the island around the time of the Irish uprising against British rule, two classics of Irish literature – Tomás Ó Criomhthain's The Islandman and Muiris Ó Súilleabháin's Twenty Years A-Growing – would probably never have been written. From these spawned a plethora of *memoirs by Islanders and their descendants, 'all drawing the last drop with melancholic longing for the past' as the Irish lexicographer (and grandson of Ó Criomhthain) Pádraig Ua Maoileoin would go on to say. It is from this collection of accounts that we get to glimpse the daily practicalities and culture of a people whose ways my generation can now scarcely believe.*

~

I'm a recovering Manchester United fan. Thankfully not many people round here know that. I grew up playing Gaelic football, and what we called soccer. Like many young boys I dreamt of becoming a professional footballer, and like many young boys I was nowhere near good enough to make it.

Due to my dad's allegiances, I began supporting Manchester United in 1983, when I was four, a few years before Norman Whiteside would win us the FA Cup and Sir Alex Ferguson's reign would begin. At the time, the club prided itself on bringing through its young players, often local Manchester lads, and had something mildly resembling a soul.

Long after the club had been bought out by American billionaires and transformed into a team of multi-million-pound prima donnas, I was still supporting them, and still addicted to football. Such was the scale of the addiction that one of the hardest thoughts I had when I was contemplating living without complex technology was the fact that I would no longer be able to watch *Match of the Day* or, considering I don't live anywhere near Manchester, ever see my team play again. To anyone who has never supported a football team, that must seem like a First World problem, and of course it is. But to anyone who has passionately followed a club from childhood, it can feel like the loss of a close friend. Now I suspect that supporting a corporate football team is a sort of toxic substitute for our basic primal need to belong to a tribe who are all bound by the same common purpose. But when one player you roared on one season signs for a rival club the next, for €90 million, the joke starts to wear thin.

Even after five months away from football, I know it's May, and that means crunch time. Excitement. Tension. Passion. Roaring. Shouting. Swearing. Cycling home from the lake one day, I notice a game on a big screen television glaring out of a pub window. I stop, stare in for a moment, and am surprised to feel absolutely nothing.

~

While I'm out in the woods climbing over an obstacle course of rotting logs and deep ditches, avoiding widow-makers, cross-cutting logs and sweating hard, Kirsty is in the cabin making dinner: a wild salad mix along with the pike and rudd I caught the night before. Are we performing traditional roles – man on wood, woman on food? It would seem so, but not consciously. We're both free to do whatever work we want, but it's just that most of the time I prefer wooding and she prefers cooking.

~

I've a thousand and one things going on in my head that I need to attend to, and they are all entirely different. Weed the vegetable garden. Water the horses. Empty the composting toilet. Mound up the spuds. Fix the guttering on the hostel. Plant out more salads. Water the brassicas. Sweep out the tool shed. Manure the pear and plum trees. Write an article about doing things like emptying composting toilets and mounding up spuds. Make a hot tub.

As I walk up the *bóithrín* to quickly water the horses, I see Tommy trundling along in the opposite direction. He stops to talk while he looks in at his field of cows. I can sense my old city mind making a reappearance. It says I don't have time to chat, I'm a very busy man. I recognise it, let it run its course, and do my best to be there in both body and mind. He tells me a bit about how a bull, when he's in a field to impregnate the cows, will take a fancy to a certain cow and, for a day or two, sit beside her. They would almost be staring into one another's eyes, he says. But as soon as the bull has 'done the deed' and spread his seed, all his interest is lost and his roving eye is caught wandering to another cow whom he hasn't yet impregnated, and he sits down beside her. That reminds me of someone, I tell him.

After he's given me sound advice on the horses, I continue up the road and bump into my neighbour JP, who is in fine form as usual. JP is in his late sixties, but you wouldn't know it. He has that kind of bubbly, youthful enthusiasm that's becoming rare even among young people. His physique, like that of my seventy-three-year-old father, belies the idea that to age is to become unfit, and he is blessed with one of those bright, cheeky faces it would be near impossible to fall out with. Whatever he is taking, I want some of it. He told me one day it was Viagra. I didn't know whether to believe him or not as he took off down the *bóithrín*, laughing his head off.

The full composting toilet momentarily pops into my mind. JP tells me that he has just got back from the Aran Islands, where he went out to visit a house he built there forty years ago, to see if the couple he built it for were still there. They were. He didn't know if they would even remember him, but they recognised him instantly, despite the years, and they spent the whole afternoon together. JP's off to a job himself, so we say our see-you-laters and I continue up the *bóithrín* to the horses.

With the horses finally watered, I meet Francie – whose field they are in – on my way back down. He asks me if we can move the horses to another field of his, as he is about to till the one they are currently grazing. No bother, I say. I tell him to give me a shout when he needs a hand with his vegetable garden. His foot has got worse, he says, 'But sure isn't it great I'm still alive.' He's eighty-four. On that note, he walks off towards his house, where he has his own horses to water.

Back at the cabin I add 'move horses' and 'collect manure' – which we'll use in the garden and orchard next year – to the jobs list. I rub out 'write article'.

~

When I first began to live without all of the distractions of modern life six months ago, I was curious to find out whether or not my overactive mind would get bored or if time would pass slower and, if it did, whether that was something I enjoyed, or something I would struggle with.

My experience has been a strange one. While the days feel more relaxed, unhurried, unstressed, the year itself feels like it is cycling through the seasons as quickly as ever. I suppose we all fill our moments with one thing or another, often forgetting to reflect on the most important question confronting us: what best to do with our precious time here?

Sometimes I feel like I'll be an old man before I solve that existential puzzle. At thirty-eight, the prospect of old age has slowly started to feel more real, like it's not just something that happens to other people. Then I look at my own mother and father, and remember that how old you are has very little to do with how many revolutions you have made around the sun, and at least as much to do with what you do with your moments in orbit.

With the exception of the odd cold day, it's much too warm to use the cabin's range for cooking and heating water. On busy days like today we have moments where we miss the ease of the gas cooker, turning a dial, hitting a button, making a quick cup of tea. It was something I took for granted for the first twenty-eight years of my life, with my generation being one of the first in the history of this island to have known no different. Sometimes I have to remind myself that I don't miss the discomfort of having to work jobs I don't enjoy to pay for gas cookers, dials, buttons and switches. Remembering is the difficult thing.

Because our rocket stove has only one hob, our outdoor meals have got to be simple. That's fine by us, as we both prefer simple food. Each of my previous ten dinners have consisted of raw salad along with boiled potatoes, cooked in the bottom pot, the steam of which cooks the vegetables on top. Tonight it's the same, with the addition of a pike caught this morning.

People in these parts eat dinner at different times of the day – Packie at two o'clock, Tommy at three o'clock, while most of my generation has it any time after six-thirty. We try to eat dinner early in the evening, as it improves sleep, but some days don't always work out like that.

As the red-hot embers in the rocket stove cool down, the still evening air is disrupted by the nasal screams of a chainsaw far off in the distance. We sit on a couple of logs eating peas, Swiss chard, mustard lettuce, runner beans, beetroot, fennel and rocket, along with the spuds and pike. I like eating each on their own, tasting their unique, individual flavours. Eventually the chainsaw stops and peace is restored, and we sit outside until the moon is high in the sky.

~

On the odd day I find myself feeling upset, or angry, or both. I don't really know why. I mean, I know why, I just don't know why it only happens on the odd day. I'll see a badger on the side of the road – big, wild and dead – or hear from a visitor about a tribe of indigenous people struggling to exist in one of the few remaining wild places of the world while resisting the pressures of oil companies, stock markets and ambition – and feel disgust at my own participation in the mechanising, homogenising, industrialising, killing culture behind it all.

I realise that by tomorrow these feelings will have passed, and I'll feel content again in this mislaid patch of Western Europe, and I don't know what to think about any of it.

And then I pick up Wendell Berry's poetry, and I find solace. In 'Manifesto: The Mad Farmer Liberation Front', he writes:

> Expect the end of the world. Laugh.
> Laughter is immeasurable. Be joyful
> though you have considered all the facts.

Laughter changes little but the quality of our days. That sounds all right to me.

~

'Mr Boyle, please explain to the class how . . .' something or other is calculated in a company balance sheet. Oh fuck. It was my first

day back after a two year sabbatical, and not only did my accountancy lecturer George Clancy – who had failed me two years earlier
– still remember my name, but within a minute he had picked my
head out of a class of two hundred. He must have taught a thousand wannabe entrepreneurs and don't-wannabe-but-will-be
bureaucrats since he had last set eyes on me. I knew he disliked
me, but I hadn't realised how much. I could hardly have blamed
him.

Thanks, George, I said, and gave him an answer that I can no
longer remember. He nodded back. I had wanted to get ahead of
myself, so a few weeks before we were due back I started to read
the recommended books on accountancy, economics and marketing. After a summer working in New York, George Clancy was a
comparative pussycat, and even accountancy seemed relatively
interesting. Relatively.

I took on a part-time job in a corner shop, earning £4 per
hour. This was 1999, the days before minimum wage and the euro
in Ireland, so in a week where I was either working or studying –
usually both – all seven days, I would earn £100.

I found most of the subjects – retail marketing, statistics, IT,
management accountancy – excessively boring, but by now I had
committed to doing it and seeing it through, and for the first time
in a long time, I wanted to do something well. The exception to
the monotony was economics, and I developed a rapport with its
lecturer. Without my knowing it, she had started the process of my
politicisation, and brought to my awareness ideas like fair trade –
a concept that was still quite radical in the early 2000s and, considering its minuscule market share compared to unfair trade, still is
today – while teaching me how to critique ideologies like capitalism, socialism and communism. Most of all, she urged us to think
for ourselves, and to question everything. Years later she would

invite me back for a guest lecture on my experiences of living without money.

My experiences of studenthood changed with my attitude towards it. Within two months I had been voted the student representative for my year. By Christmas George Clancy was stopping to give me a lift into class in the morning, and it was he who handed me a First Class Honours degree at my graduation as he shook his head and smiled. But by then I didn't give a shit about the piece of paper, or the result.

After four years of studying business, one thing stuck with me: in all of that time, not once did the word 'ecology' get mentioned. Even back then I found that odd. How could I claim to understand economics when I knew nothing of the natural world on which all economies ultimately depend?

I had done what I set out to do, but I knew I needed to change something important in my life, and not just the scenery; for no matter where I would go, I knew I'd always find myself there. I had been wanting to give up booze for a while and become vegetarian, neither of which I imagined would be easy to do while hanging out with the groups of friends I had in Galway. They were only vegetarian when they were sleeping. Come to think of it, most of them were only sober when they were sleeping too. I packed my bags, said my farewells and, like so many of my ancestors, I took off for foreign lands. I didn't know if I would ever come back to live in Ireland.

I found a flat in Edinburgh within days, and a few weeks later I started a job at the checkout in a supermarket. The work was simple, and it paid the bills – just about – but reading manuals detailing how we had to greet customers and wear our clothes soon started to pall, as did having to offer plastic bags to every customer who bought a sandwich or a soft drink. In the little corner shop I had worked in, I knew most of the customers by name, what cigarettes they smoked, the newspaper they liked to read on any given day. I almost knew their lotto numbers. Now I was being told I had to say things like 'How can I help?' and 'Thank you for shopping with us' instead of what would naturally come out of my mouth.

I bought my own weekly groceries from a rival supermarket. The food there was cheaper. Back then I still assumed cheap to be a good thing, not thinking that my gain may be considered a loss to somebody (farmers, factory workers) or something (land, insects, animals, fish, rivers, forests, oceans) else. One week I picked up a plastic packet of South American tofu which was labelled 'organic'. I hadn't a clue what that meant, so I read the story on the back of the packet. Having been versed in marketing, I assumed that most of it was exactly that – a story – yet it got me thinking. Since I was a child I'd had a fondness for the natural world, but it wasn't until this moment, standing opposite an aisle-length fridge in a multi-national supermarket, that I realised I wanted to start engaging with it.

Within two weeks I had quit the supermarket job and was working as a logistics manager – a storeman – for an organic food company. It was here that I would learn how little I knew about real economics.

~

As the pike lies dead on the butcher's block, fierce and beautiful and ancient and strong, I remember the moment I killed him. He was flipping around on the grass, tired but still fighting hard, his eye – the one that I could see – fully alive with none of the fear you might expect. Of all the wild creatures of Ireland, the pike is one of those I'm least concerned about killing. Still, taking the life of another creature – especially one so wild and free as this – should only ever be done with the reluctance of one who needs to eat.

The pike had displayed no such civilised sentimentality as he pounced on what he thought was another fish at the end of my invisible line. A fatal error of judgement. Instead of getting food, he was becoming food. I tried to keep this in mind as I held him on the cold stone, before whacking him over the head – once, twice, three times – until his eye opened up with that enlightened look of one who suddenly understands something no living creature knows, or can ever know.

Squeezing the underside of the pike gently, the remains of his previous dinners squidge their way out of his anal opening, from which I slit him all the way to the gills. His internal organs come out easily. I feel something hard in his intestines and, curious, I cut them open to discover a small piece of gravel. These guts will be an offering to the local wildlife later.

With two angled cuts behind the gills, off comes the head, before I work my way through the fins. All of this offal goes into a pot, along with his innards and the scales I've scraped off, and will soon be cooked on the rocket stove before being left to stew in a hay box, where it will slowly become chowder. Brains, bones, fins, skin, heart, eyes – potent stuff. I cut his body into steaks, keeping the liver for the pan – there will be more than enough for myself and the five others who live on the smallholding.

The pike soup is thick. I'm no scientist, and have no desire to be, but as I drink the first cup my body tells me it is packed with nutrition and goodness. It couldn't not be. In the old times this bone and offal soup would be passed around from neighbour to neighbour, each boiling it up and getting a turn out of it. But those were the days before easy come, easy go. Now even some of my friends who are local food advocates won't eat it. Too fishy, they say.

~

For the Great Blasket Islanders, living off the land and the open sea surrounding them was not the lifestyle choice that some might, understandably, argue it is for me; though I, for my part, no longer consider it a choice. Not really. No, for the Islanders it was hard economic reality, and something ingrained in the sinews of their culture.

Aldo Leopold once wrote that one of the 'spiritual dangers' of not spending time on a small farm was that you may 'suppose that breakfast comes from the grocery'. In order to avoid such danger, he said, 'one should plant a garden, preferably where there is no grocer to confuse the issue'. These Islanders were in no such danger. Their nearest shops were in Dingle, a 5-kilometre row and 20-kilometre walk away; that is, if the ocean was even passable, which it often wasn't. Therefore it was critical for them to make full use of what grew around them, and not to be dependent on the vagaries of the Dingle market, or the money they seldom had.

They ate a natural, balanced diet, and were said to be among the healthiest people in Ireland at the time. They had to be, for there was no doctor or nurse on the island either. Potatoes were a staple which everyone grew. The famine hit their crop just as hard as on the mainland, but because they had a more varied diet – and the skill to attain that diet – they fared much better than those in Ireland's towns and cities.

On top of this they grew oats and rye, and would sometimes buy large sacks of flour in Dingle with the money they received from selling mackerel. Other vegetables were not a major part of their diet, probably because the weather there wouldn't have been conducive to good growth. Instead they would eat seaweed – sea-belt, murlins, dillisk and sea-lettuce. They would dry dillisk on their roofs, before making it into a chew. Seaweed was abundant, though tough work to collect and haul up the steep hill to their village. Along with soot from their chimneys and mussel shells from the nearby island of Beiginis, they also used seaweed to fertilise their potato ridges, and so they effectively ate it by way of their spuds too.

Seafood of all sorts was a major part of their diet. They collected limpets and periwinkles down at the shore. In the summer they would eat fresh mackerel, roasted on the tongs. With the exception of that and bream, they preferred to boil fish for their dinner. They had no taste for salmon and so, according to Pádraig Ua Maoileoin, whenever they caught one they would either throw it back or cut it up as lobster bait. Nowadays a single, good-sized salmon can fetch in excess of €100.

In the winter they often ate cured meat, which they had earlier hung over the mantelpiece to dry. Those not keen on fish for breakfast might have an egg or two from their hens, who were known to make a mess of the houses' reed-thatched roofs, within which they would often be caught nesting. One time, while Tomás Ó Criomhthain's father was having dinner, a young chick fell from the roof into his mug of milk.

Others would have their own oats for breakfast. Those who had a cow – most people – might have it with a glass of thick, sour, unpasteurised milk. The same cow would give them butter, and buttermilk on churning day. In order to keep the supply of calves – and thus milk – steady, they had to row a cow to the mainland on one of their canvas-clad boats, where she would be serviced by a bull. This was a tough, dangerous job that would often take a day. Once, a cow drove her horn through the canvas, almost drowning all of the men on board. Eventually they were awarded a free bull from the Kerry County Council, which they took out to the island so that he could impregnate the cows one by one where they were. The first year, after he had seen to them all, a few of the Islanders took him off to the small neighbouring island of Beiginis, and out of harm's way. The next morning they found him back on the island; the horny fellow must have swum through the night to get back to his harem.

Meat naturally played a less important role in their diet than fish. Seal meat – which there was plenty of, though dangerous to hunt – was highly valued, and it was easy to barter for its same weight in pork when they went to the markets in Dingle. The skins could fetch £80 alone in Ó Criomhthain's youth. By the time he was old, however, the taste for seal meat had gone. Most families killed a sheep twice a year, some of which would be eaten fresh, the rest cured. You were allowed to graze twenty-five sheep on the island for every one cow you owned, and the minimum most families kept was 'a sheep to sell, a sheep to shear and a sheep to eat'.

The islands were teeming with rabbits. Muiris Ó Súilleabháin and his young friends would hunt them with ferrets or snares when they weren't stealing seagulls' eggs on Beiginis. Pádraig Ó Catháin – the man who would become the King in a place where Ó Criomhthain remarked that 'kings weren't so hard to satisfy' – caught a dozen rabbits one day when the new island school had closed because the teacher had died. Their parents would have caught seabirds – young gannets, puffins, storm petrels, razorbills – at any given opportunity, and roasted them in the pot oven on the fire.

It was this diet that kept them well. There was no industrial healthcare system at the time – no dialysis machines, no stents, no replacement hips, no ambulance to come and save them if they got ill. Whenever they did get sick they mostly relied on home remedies. In his memoir, Michael Carney recalls how, after breaking his leg, his father took him by boat, over the rocky waters of the Sound, to a bonesetter – a local farmer – in Dunquin, who duly set it back in place without anaesthetic. His father was also a rudimentary sort of dentist, using pliers and pieces of string tied to the door as his methods of extraction.

The women would usually fetch the water from the well, which was located in the upper village, itself not even a stone's throw away from the lower village. This, often to the men's dismay, was also the scene of much gossiping. If they were working in the fields, the Islanders would usually mix this water with a dash of milk for extra sustenance. They knew nothing of tea until a tea chest washed up on their shores one day. At first they used it sparingly at Christmas, with 'the remnant saved up till the next Christmas'. But before long they were having it every day, and its introduction seems to have changed the entire eating habits of the Islanders. Instead of having two meals a day – a sturdy breakfast and dinner – they began having four small meals, two of which consisted only of tea and bread, what Ó Criomhthain lamented as 'a miserable bite or two'.

They managed their diets of vegetables, eggs, milk, meat and fish without a fridge or freezer, or electricity for that matter. They cooked without gas or oil. It wasn't an easy life, but then none of them ever grew up expecting one. I remember M. Scott Peck writing, at the beginning of A Road Less Travelled, *that once you stop expecting life to be easy, life suddenly becomes a lot easier. This was true for the Islanders. Considering the scale of anti-depressant use in contemporary society, it appears that life in the industrial world isn't easy either.*

Looking back from our vantage point, my generation would consider the Islanders' way of life extreme, though to them it was normal. We'll never know what they, if they could have looked forward, would have thought of our world.

~

The pub is packed. Kirsty is lured in by the sound of John O'Halloran on the melodeon intermingled with raucous, melodious laughter. As we walk in the door the scene reminds me of Kavanagh's 'Inniskeen Road: July Evening':

> There's a dance in Billy Brennan's barn tonight,
> And there's the half-talk code of mysteries
> And the wink-and-elbow language of delight.

It's an older crowd, but there's a young couple, I guess somewhere in their twenties, sitting below us at a table to our left. He is scrolling down his smartphone, whose glare catches my eye among the darkness of the pub. She is trying to hold his hand, but he doesn't appear to notice her gentle attempts. The philosopher Alain de Botton has said that 'True love is a lack of desire to check one's smartphone in another's presence.'

By now Kirsty is out on the floor, tapping the steps of the traditional *sean-nós* dancing – a much older, more fluid, less rigid form of Irish dancing than the modern style popularised by *Riverdance* – to her friend O'Halloran's reels. It's a wonderful thing to watch her dance, not so much for the spectacle – a fine thing in itself – as for the fact she is most alive and radiant when she is dancing.

O'Halloran has finished his third reel when I notice, out of the corner of my eye, that the young woman is still trying to hold her partner's hand. He's checking the football scores. Part of me, momentarily, wants to ask him who won, but I catch myself. She heads for the toilet as Kirsty makes for the floor again.

~

When I turned the light switch off, for good, on winter solstice, the natural light was either dim or dark for sixteen hours out of the twenty-four. On those days when the sun struggles to raise itself above the conifers to the south, we could easily get through two or three beeswax candles a day. Almost six months later, we use none.

Because of energy policy in Ireland, electricity consumers pay a hefty standing charge, on top of which a small per unit cost is applied. I've no doubt there's plenty of sound financial logic behind it, but at a time when the world's scientific community is pleading with governments to reduce their emissions, policies like this effectively mean that there is little or no financial incentive for people to minimise their energy usage, as the cost of actually using the energy waiting behind the switch is so marginal. Reduce the standing charge to almost zero, and hike the cost of the electricity so high that carelessness matters, and unused lights, computer screens and devices would go off overnight, quite literally. At the moment you still have to pay the charge, even if you don't use a single unit in June. It all reminds me of an old Irish proverb we'll likely learn when it is too late: *taréis a tuigtear gach beart.* 'We learn when it is too late'.

Reading Barry Lopez's *Arctic Dreams*, I become aware of how our conception of time, and what a day is, belongs to a place as much as the physical life which more obviously comprises it. For the life of an Eskimo in the Arctic, 'The idea that the "sun rises in the east and sets in the west" simply does not apply,' and I come to understand that the 'thought that a "day" consists of a morning and a forenoon, afternoon and an evening, is a convention, one so imbedded in us that we hardly think about it'. If I stood at the North Pole on summer solstice, I would see the sun make a 'flat 360° orbit exactly 23.5° above the horizon'. Lopez adds:

> In the Temperate Zone, periods of twilight are a daily phenomenon, morning and evening. In the Far North they are (also) a seasonal phenomenon, continuous through a day, day after day, as the sun wanes in the fall and waxes in the spring. In the Temperate Zone each day is noticeably shorter in winter and longer in summer but, still, each day has a discernible dawn, a protracted 'first light' that suggests new beginnings. In the Far North the day does not start over again every day.

With my way of life, preparation is critical. Just as November is not the time to be getting your winter's wood in, I start thinking of darkness in June. With that in mind, I go out to the potato field to cut rush. Its pith, which has all the properties needed for an effective candle wick, will help enlighten my winter, and it doesn't levy a standing charge for standing in the field in June.

~

We dug out a pond with spades when we first got here. It took two long days immersed in mud up to our arses, our wellies absolutely submerged and utterly pointless. We may as well have gone in wearing slippers. The idea was to create a habitat for a range of species that would be beneficial to our vegetable gardens and potato field, and the landscape as a whole.

Looking at the frogspawn next to the bank now, I'm reminded of how simple it is to bring life back to a place. It seems that all we really need to do is provide a habitat that is protected from our machines and our need to control, give it a little time, and nature – often mysteriously – does the rest.

I'm struck by how calm and at peace with things the pond is today. There's a steady trickle of fresh water coming in from the west, heading out towards the east. Perfect. If no new water at all were to flow into the pond, it wouldn't take long for it to become stagnant. Too much water too quickly and its stability can become overwhelmed, eventually filling up with the silt that was washed away upstream.

My only job then, as pond steward, is to make sure I keep the balance right.

Summer

Ours is essentially a tragic age, so we refuse to take it tragically.
The cataclysm has happened, we are among the ruins, we start
to build up new little habitats, to have new little hopes. It is
rather hard work: there is now no smooth road into the future:
but we go round, or scramble over the obstacles. We've got to
live, no matter how many skies have fallen.

D.H. Lawrence, *Lady Chatterley's Lover* (1928)

Today is the longest day of the year. Summer solstice. When I
lived in the city, working the 7–6, each day felt as long as the next.
Electric light, alarm clocks and closing hours standardised my
experience of the seasons. At the heart of modern society lies not
just capital, fossil fuels and ambition, but Greenwich Mean Time
– which, in *Pip Pip*, Jay Griffiths calls the meanest time of all.

I awake, not knowing what time it is. This has become normal. As
the soft, red morning light drifts in through the open window above
my head, I find myself not caring what time it is, either. A privileged
position to be in, it could be said, but while I thank the gods most
days, as a couple we live on a fraction of what is considered to be the
poverty line for a single person here in Ireland, and without electricity
or running water. So not most people's idea of privilege.

Whatever time it is, it must be early. As I slowly come around
I feel refreshed, and it feels good, natural, life-affirming to wake up

with the light. I was in bed by last light last night, and so considering we get about six hours of darkness at this time of year, that's good sleep for me. In fact, it's longer than when I used blackout blinds. The quality of rest feels better, too.

Six months ago, such an outcome would have seemed a minor miracle. Now I've started to take a good night's sleep for granted. I don't like taking things for granted.

~

There are seven heaps in our compost bay, and six of them are full. When I say full, I mean half-full, as at this time of year they've shrunk from the full pallet height to the midway mark. That's good, as it means the elements and thermophilic bacteria are doing their job. But it also means that my first job this morning is to turn two heaps over into one, for a couple of reasons. One is to make space. The other is to reintroduce air into it, and thus aid decomposition. Most gardeners and smallholders don't bother turning it, and it's not essential, but I find that it's worth the effort.

As we use a 'humanure' system here, which incorporates human piss and shit into the mix, there's a part of everyone who lives here, and a few of the visitors, in the heaps in front of me. Most people, having never done it, find the thought of turning this kind of compost disgusting, but that's just one way of looking at it. In it I see stories and memories and history, and a great link between a place and its people. All I am really doing is making soil, and that seems to me as good a way to start the day as any. And in doing so I'm continually reminded that the boundaries between us and the land which nourishes us are nowhere near as clear as we might like to imagine.

~

A friend, who had travelled the Indian subcontinent for six months, tells me about a small village-worth of women, old and young, whom he once met on the banks of a river in remotest Pakistan. They were washing clothes together. Not having a shared language, he didn't understand a word of what they were saying, yet was struck by how much they appeared to be enjoying themselves – laughing, talking incessantly, being playful – doing something, he said, he would hate to have to do himself.

My own past experience of clothes washing bore no resemblance to such a scene. In an individualising, atomising Ireland, my own experience of everything bore no resemblance to such a scene. For most of my life, washing clothes meant loading up a drum, turning a dial, hitting a button, and going off and doing something else. But having given up on all things automatic seven months ago, this had to change. If I'd never seen a washing machine before, or had grown up in a remote region of Pakistan, hand-washing my own clothes would simply be an essential fact of life that was entirely unworthy of mention. But I have seen a washing machine, I know exactly how quick and efficient they are, and I was brought up by a generation who were only too keen to swap the hardship of hand-washing for the flick of a switch. So I accept that I'm probably never going to enjoy it, especially as long as it remains a lonely, isolated process far removed from any social ritual.

With this in mind, I've become frugal with my clothes usage of late. At this time of year I spend a lot of my time in shorts. I find that wearing too many clothes in summer can soften you up too much for the harder months of winter. So the wash basket tends to fill up only once a month.

Clothes washing morning, first light. I gather some dry twigs and get the rocket stove fired up. On top of it I rest an old, blackened pot which I fill to the brim with spring water and chopped-up soapwort (*Saponaria officinalis*), a perennial plant we grow on our smallholding. Soapwort contains saponins, and works just as well on your body and hair as it does on your clothes. Historically its use was conventional, and you can still find it growing wild around the sites of old Roman baths.

I heat up the pot until the water inside is very warm, taking care not to allow it to boil, as that would kill the active ingredient in the plant. The time taken to make it probably doesn't compare favourably with the fluorescent green stuff you can buy at the supermarket, but getting the best financial value for each moment in my life hasn't been my primary motive for a long time.

Next I soak the clothes in cold water in the washing basin, while I pour the pot of soapwort liquid into a small hand-crank tumbler. I wring out the cold water, put the clothes into the tumbler, and spin it around by hand for about ten minutes. As I do so, I ponder how much harder this would be for a family of six, and appreciate why working parents love their washing machines today. At the same time, I can appreciate why such a modern perspective is part of a much bigger economic, cultural and ecological problem.

Once the spinning is done, the clothes go back into the basin where they are first scrubbed, and then rinsed and rinsed, until the water remains clear. From there they go through the mangle and, at this time of year, onto the line in the garden (in winter they would be hung on the drying rack, which Kirsty made from hazel rods, above the range in the cabin). The whole process takes most of the morning, and by the time it's done I'm glad it's over for another month.

As I'm putting the mangle away, my mind wanders back to Pakistan, and I contemplate what I and my progressive culture are losing in our 'progressing'. I'm reminded of an experience of my own, somewhere I can't quite place on the island of Java, in the Indonesian archipelago. I was with two friends, Gavin and Nigel – both of whom kept themselves in good shape – hiking up a steep mountain during a summer spent exploring the country. After a few hundred metres of sharp incline in blistering heat, we stopped for a quick breather. As we were standing there, taking in the view and preparing ourselves for the next part of the climb, a young girl – who looked about seven or eight years old – walked up behind us with a full bucket of water hanging off each arm. She smiled at us, and said hello. We looked sheepishly at each other, and followed her up. We thought about asking if we could carry them for her, but she didn't seem the slightest bit put out by the buckets and, quite frankly, she looked more capable than any of us.

A few hundred metres further on we reached a small cluster of huts, where a handful of men and women were washing their clothes. Funnily enough, at the time I remember thinking how primitive their lifestyles were. Now I wish I could go back and learn from them.

~

Aside from my own experience, most of what I know about pike fishing I've gleaned from Maureen, the owner of the nearest fishing tackle shop 20 kilometres away in Portumna. I remember the first time I went in there, looking for what was, in hindsight, an unnecessary bit of kit, and being talked out of buying it by her. If you're really enjoying fishing in three months' time, she said, come back and I'll sort you out then. She was right. I ended up not wanting or needing it, though not for a lack of interest in fishing.

I pop in to say hello. She tells me that she's struggling to survive, though she shows no sign of relenting. Online stores, with their warehouses in industrial estates, often in cheaper countries, are killing places like hers, she says. It's not a surprise. I have never seen a single website that has talked anyone out of buying anything it sells. And it's hardly for the want of unnecessary stuff on the internet.

~

I've got to be up early in the morning. I usually am anyway, but this time I have to be somewhere important. Somewhere civilised. I have no alarm clock, so I need to trust myself which, after a lifetime of putting my faith in our contemporary religion, Technology, is not as easy as it ought to be. But I do it anyway.

I wake up. It's early. The people I meet, for whom I'm in good time, tell me it was my body clock, but I'm not a machine made of cogs and springs. I'm an animal, made of feelings and failings, hopes and flaws, instincts and intuitions. I don't know what explanation the priests – The Scientists – of this new religion would give but it wasn't the clicking of some Cartesian gear which awoke me. No, I prefer to think of it as a knowing beyond knowing.

~

Many years ago, as we walked into the woods on the first morning of a bushcraft course, our teacher – my friend Malcolm Handoll – asked us all if we could take off our boots and wellies, and continue barefoot instead. Wide-eyed and tentative, we looked at him and each other – really? – as one of our small group asked Malcolm why. After all, it was the middle of November.

Malcolm's response has stuck with me ever since. He told us to 'imagine a life wearing boxing gloves on your hands. That is how my feet feel in wellies.' As we unlaced our boots and reluctantly pulled off our wellies, he implored us to think about how it might feel not to be able to touch anything, ever, with our naked hands because they were always protected by a pair of thick, rigid gloves. I told him I had never looked at it that way before. Okay, he said, leave your boots there and come follow me.

Being a frosty late autumn morning, my first concern was that my feet would get so cold I wouldn't be able to enjoy the course. At first they did, but as soon as we got moving I was surprised to find them warming up. I could feel the blood flowing, and the nerves in my feet tingled, as if they were keen to explore. It felt like a foot massage and reflexology session rolled into one and, for the first time in my memory, they felt alive and connected to the great living, breathing, wild beast below them.

My second concern was that my feet were going to get injured, but that never happened either. I found myself paying careful attention to every footstep, stepping over jagged stones and thorny plants, and noticing things I suspect I would otherwise have missed. I walked more sensitively, more consciously, not just trampling over things with the disregard that a common technology such as boots allows. All of a sudden I found myself alert to every crunch and crack and texture under my soles, for no other reason than because I had to. This wasn't connection with the earth for spiritual reasons, this was connection with the earth for physical, practical purposes. Or, who knows, maybe it was both.

When the course was finished I put my boots back on, and kept them on until now. Samuel Beckett was right. 'Habit is a great deadener.'

The air last night was fresh and crisp, the sky utterly cloudless, the kind that punctuates two perfectly blue-skied summer days. The sun has only just snuck up over the eastern horizon, where it looks like an ember left over from last night's campfire. My first job of the day is to water the plants. On my way out the door I catch myself, take off my boots and socks, and leave the house barefoot for the first time since I last saw Malcolm.

Instantly I feel the soft dew awaken the insides of my toes. They're cold to begin with, no doubt; it's not long ago they were under a blanket. But just as your body reacts to plunging into a lake on a summer's evening, they soon acclimatise to the new conditions. As I walk towards the potting shed I notice new plantain leaves coming through the grass, and droplets of dew on spider webs (how many thousands of spider webs have I mindlessly demolished before?). Just as my mind wanders to the day ahead, as it has a tendency to do, I snag my toe on a sharp, protruding rock. It hurts, and with the pain comes an important lesson in mindfulness.

Coming onto the *bóithrín*, as I go to collect water from the spring, I see that my eyes are fixed on the road, and not the wildlife around me. I feel every bit of gravel and stone, my soft feet wincing with every second or third step. They need to harden up. They're certainly not loving it, but it feels fine nonetheless. My feet seem to know their way onto the strip of green grass up the middle of the road, and the world around me feels gentle and alive and breathing again.

As I turn the corner towards the spring, I find myself hoping that Kathleen is not up and about yet, for if she sees me coming in bare feet at this hour of the morning she'll think I've finally gone mad.

~

Kirsty is struggling. She has a natural love for the natural world – or at least a kind of hatred of what we are doing to it in the name of 'progress' – but it is still, in some small part, an intellectual love, one that hasn't fully percolated down to the marrow of her bones yet. Such things, for our disjointed generation, take time. A lot of time. She understands the ecological imperative – and thus the social imperative – of radically changing the way we relate to all that lives, perhaps more than anyone else I've ever met. She feels it too. Too much sometimes. But what she truly loves is dancing. And what she loves more than dancing is dancing to live music. You should see her.

She's a sociable creature – aren't most of us to some degree? – and so I know she feels a sense of isolation out here at times. It's a close-knit community of people, but the young have followed those who jumped ship before them to the cities, and so it lacks that youthful spirit that every place needs to thrive. We're trying to bring it back to this place, but it's a long and sometimes lonely road. Having no car, and public transport in rural Ireland being what it is, it's not easy for us to get to the places that are culture rich, nature poor.

Today I find her quite upset and alone. It breaks my heart to see, partially because I love her like nothing else in the world, even if I'm not always aware enough to show it, and partially because I know the feeling myself. This can seem like a hard way of life until that love for all of life – and not just human society – trickles down from the head and into the veins and bones. It being natural, our need for loving human companionship never leaves us, nor does our longing for a sense of belonging. It has taken me over ten years to feel, as Wendell Berry once wrote, 'the peace of wild things', and even at that I have many moments when I miss what I once knew well. She's not had that time yet. And being naturally nomadic – perhaps we all are, as is seen in the modern trend to fly all over the world, to explore one thing or escape another, that has become a toxic substitute for real journeying – it crosses my mind that maybe this rooted life will never be for her, and we will have to find a way of weaving the adventure of nomadic life into it.

Not knowing what else to do, I hug her. I tell her to put on her favourite dress, to grab her hula-hoops and tap shoes (she's becoming an accomplished *sean-nós* dancer) and that we're going to find some music. She has been gently asking me to relearn the tin whistle for over a year now, so that we could play and practise together in the evenings, but I've made one excuse or another and now I feel a bit ashamed.

As we enter the pub, everyone comes over to ask if she'll do some dancing. The hula-hoop is still a great novelty in these parts. Everyone around here loves her, as people do anyone who is doing what they love. As she dances the steps to 'Cooley's Reel' and 'Drowsy Maggie' she is reborn and rejuvenated.

I know I need to make more time for this, and for her. I tell myself I've been busy, but it's more a case of getting my priorities wrong. Again. I know this now, and I hope I haven't left it too late.

~

A friend of mine, who runs an organic farm nearby, calls over with 22 kilogrammes of blackcurrants in a couple of crates. He tells us that he planted lots of blackcurrants years ago, but because people have stopped making jam and other preserves, he no longer has a market for the produce.

We're only too happy to take them. A quick calculation tells me that they'll make 75 litres of wine and twenty jars of jam, with enough left over for breakfast for the next week. There's a day and a half's work in it, but that should keep everyone in booze and jam for the winter coming, and we still have blackberries and raspberries to come.

I offer to buy the currants from him – I know only too well the amount of work involved – but he won't hear of it, and so we settle on as much wine as he likes, instead.

~

A neighbour of mine asks me if I'll put the credit he has just bought at the post office on his phone. He says he has forgotten his reading glasses, but I know that he can't read. He told me so one night after a few drinks. It's funny how, in these times, we're embarrassed by any lack of intellectual skills, yet we seem perfectly fine with our insufficiency in the most basic of life skills, such as feeding and housing ourselves. The same neighbour knows things about horses that I will never know.

It's no time for ideology, so I take the phone from him. For the first time in nine months I tap numbers into his phone, which allows me to tap in some other numbers from the voucher which he bought with a bunch of other numbers. The computer-generated 'voice' tells me he's all topped up. He can now, almost instantly, talk to people on the other end of the world, and do many other things I'm no longer able to do.

~

The gooseberries are out, and almost ready for picking – but not quite. Timing is everything. Too early and they are too bitter and small. Too late and the birds will have them all before you even know about it. When I lived without money I emerged from my caravan one morning to see thirteen grey squirrels gathering walnuts in the nuttery. Before that I had only ever seen one at a time. They had known exactly when to come, and they had got there just before me. In nature those who thrive are those most in tune with their landscape, those most alive to the rhythm and pulse of life. It is not so much survival of the fittest as survival of those that fit in.

I pick the gooseberries at dawn, most of which will be for breakfast. The birds have already had the first of them. I take what I need for now, and leave the rest to ripen for myself and the birds, whose ways and song are sustenance for the soul. As I sit and jot down these words in the garden, a blackbird forages from what I have left behind.

~

The pencil is in my hand. I remember holding it, back in January, and thinking, 'This is fucking impossible.' I had just spent three weeks attempting to write without the aid of copy-and-paste, delete, spell-check, the World Wide Web and all of the research and editing tools that a flat, rectangular piece of plastic once afforded me. Before, writing had been relatively easy; I would blurt out my thoughts onto a screen, which I would then make sense of using Microsoft Word. Quitting computers and the internet, however, made it clear to me that, for all of my adult life, I had been a cyborg-writer, the words on my pages the result of something part man, part machine. My previous books would certainly have been a lot different, for better or for worse, if I hadn't used a computer to write them. Actually, they probably wouldn't have been written at all.

In the heart of winter I noticed myself scrunching up piece of paper after piece of paper, each one containing too many errors of grammar, type or judgement, or needing more editing than one sliver of wood pulp could endure from a rubber. I had been getting accurate at throwing these balls of paper into the wood basket, and I consoled myself that at least my day's work was useful for starting the following day's fire. The pads of paper I used were cheap – much too cheap. If I'd had to make my own paper, like my friend Fergus Drennan did for *The Foraged Book Project*, there was no way I could have been so extravagantly wasteful with it. My plan had long been to start making my own mushroom paper, but I knew I needed to relearn how to write first, otherwise making paper would become a full-time job in itself.

By April, there were some signs of progress and hope. Writing was still bloody challenging. But impossible? I wasn't so sure any more.

It's now July, and I'm halfway through this book. For the first time in my life I'm actually enjoying the process of writing. My head no longer hurts at the end of a long day. I find myself staring into the orchard for prolonged periods before I even put pen to paper, but when I finally act I can write fifteen hundred words without stopping. My thinking has got slower. Just as carpenters always recommend measuring twice and cutting once, I've begun thinking twice and writing once.

What comes out may be as stupid and shoddy and humanly imperfect as ever, but at least they're my own thoughts in my own order. As the way we live and think is shaped by the technologies we use, writing without the aid of computers feels important to me. For as Sydney J. Harris once said, 'the real danger is not that computers will begin to think like men, but that men will begin to think like computers.'

~

While the rest of Knockmoyle is sleeping restfully, the irrepressible JP and I are at large. He's working away two fields up from where I'm collecting horse shit for our vegetable garden. I throw him a wave, and before I know it he's in the field next to me. He tells me that he was meant to be on the road half an hour ago, but he doesn't seem the slightest bit bothered about it. As well as being a farmer, JP is also a carpenter and a general builder, and like everyone else around here he uses his skills to supplement his farming.

Before he goes off to work on his daughter's new house, he tells me some of the latest news. The banks, he says, have given out twenty thousand mortgages, despite there only being ten thousand houses on the market. He says it's all going to go 'cracked crazy mad' at some point again. *Taréis a tuigtear gach beart.*

In the same breath he adds that the US government has decided to reduce its corporation tax rate and that, because the big American technology companies are mostly only in the country for its 'competitive' tax rates, our politicians are falling over themselves to find ways of keeping the likes of Facebook, Google and Apple from leaving Ireland and going back to where they came from. We shed blood and countless tears to win our political independence, it seems, only to give away our economic independence – our self-reliance as a people – without so much as a murmur. He says that we've always been influenced by America, and that its sway over us is only getting stronger.

As he is walking away, he tells me that when he was a young man he heard rumours that Americans had nursing homes where people would put their elderly parents while they tried to get ahead, stay ahead or simply stay afloat. He says he remembers thinking, at the time, 'Isn't that a terrible country that would do that to its old people?' And look at us now, he says.

See you later, Mark, isn't it a great morning to be alive and have your health. And with that he is gone, and I go back to picking up fresh shit from the field.

~

Science is a lot like my father. When I was growing up he would be forever taking apart things that were already working fine, more out of curiosity than anything else. He could then tell you how everything worked. It was interesting. Unless you were my mother. It would drive her mad. More often than not, whenever Dad went to put the radio or telephone he had been tinkering with back together, he would realise that he had lost a few components, sometimes important ones that weren't easy to replace. For a while, he would claim that it didn't need that part anyway, until it would find its way onto a shelf in the shed, no longer able to work.

~

Almost-empty glasses dot the round, wooden tables of the *sibín*, and signs of a good night linger everywhere – guitars, mandolins, banjos, fiddles, tin whistles and *bodhráns* are strewn across the bar and chairs, embers are still glowing in the fire, and the smell of stale homebrew fills the air.

I pour the bilge water from the glasses into a bucket and make for the vegetable garden, where I decant it into small bowls, each of which has now become a slug trap. Like most of those waking up in the hostel, slugs suffer from a fondness for beer; but while excess has certainly ruined many a good person or family, one drink can prove fatal to slugs.

The next morning I examine the bowls. Carnage. There's six in one bowl, four in another, two in the next, and on and on. It's highly effective. Too effective for my liking. My civilised mind kicks in and I start to question the ethics of it all. On one hand, I have to eat, and slugs are only too proficient at finding precisely the plants that I've developed a taste for. On the other hand, I'm knowingly drowning – albeit in beer – other sentient creatures for the crime of enjoying those greens which, by way of having planted them, I consider mine. I'm not sure what exactly, but something about it doesn't sit right with me.

I pick each slug out of the beer bowls and leave them to dry out on the wooden frame of the raised bed. For a moment I consider eating them, but when I go back to where I left them they're gone. Either they've woken up with a frightening hangover, and have gone to hide somewhere dark, or there's a couple of wild creatures trying to get merry in the hedgerow.

~

It wasn't exactly every schoolboy's dream job – do schoolboys dream about any job? – yet I found a strange sense of purpose the moment I started working for a small, independent organic food company in Edinburgh around 2002. Unlike my previous supermarket job, the people who worked there were passionate about selling real food that was good for people, booze that was fairly traded and, like me at the time, the shop was entirely vegetarian. And, though my perspective would change quite dramatically five years later, at that moment I felt I was part of something bigger than myself, something that mattered.

The enthusiasm of the people who worked there quickly rubbed off on me. I was working almost as hard as I had in New York, but I was learning a lot; less about myself and more about the political world around me. Part of my job was to build relationships with local, organic producers of food and drink. I found myself learning more from them about economics than I did in four years of a business degree. Every day I would hear a little about what they did for a living, and why they did it, along with some of their stories and the difficulties they encountered in a world going in the opposite direction. I found myself admiring their belief, their patience, their tenacity to do it anyway.

Beekeepers would teach me about the fundamental importance of pollinators, the uses of beeswax and impact of neonicotinoids on honeybee colonies. I would learn from salad growers how pesticides and chemical fertilisers were polluting the soil and eroding biodiversity, making rivers and streams undrinkable for people and uninhabitable for aquatic species. Local chicken farmers, operating on a small scale, would tell me about the conditions that factory-farmed – and even so-called free range – hens would have to live in. Before I knew it I was reading books by Noam Chomsky, Naomi Klein, Jared Diamond and Vandana Shiva. Work was no longer just work, it felt like a political act.

Through working all the hours God sent, I managed to pay off the student debts I had acquired despite having worked every hour God sent in the little corner shop in Galway. But after two years living in Edinburgh, I still didn't feel at home. It was clearly an affluent city – manicured, controlled, sanitised – and yet homelessness was rife. I was becoming increasingly drawn towards activism – something, anything – but I couldn't find any movements for change there. As long as wages are high, desire for change is low.

Together with my first love, a Finnish girl called Mari, I decided to move to Bristol, where I was told that all of the action happened. We got ourselves a second-storey flat directly opposite a busy motorway on stilts, under which we would see prostitutes give blow-jobs to drunk men early on Sunday mornings. The noise was relentless and I would often worry about Mari if she was out at night alone.

Two weeks after landing in Bristol I went for a part-time job with another organic food company which had a supermarket, café, box scheme and walled garden. I wasn't impressed by the place on first sight, or how it was run, and I didn't hold back in telling the owner so in the interview. I left assuming I would neither get nor take the job. Five minutes later I got a call from the owner, asking me if I wanted to manage the company instead. Surprisingly excited by the challenge, I said yes.

Within a year, and with a lot of graft and risk, we managed to turn the place around. I was a hard boss, driven by the fact that I knew that good people – growers, beekeepers et al. – were struggling, and I didn't want to see them struggle so much any more. It was my job to make sure local people bought the food these producers made and not something that was abused, somewhere on the other side of the world, by some faceless corporation. But one day I walked out of the office, down one of the aisles, and suddenly found myself frozen still. All I could see was wall-to-wall plastic. Cacao nibs in plastic packets, vitamin pills in plastic tubs, water in plastic bottles. Bananas from the Dominican Republic, sweet potatoes from Israel, mangoes from somewhere else not very near. For three years I had been ranting and raving about sustainability, but it only hit me there and then that even this organic industry was five thousand miles from being sustainable. And even if it was possible, why would I want to sustain a plastic culture anyway? It wasn't even sustainable for me. I was working over sixty hours a week (and pushing our own staff too hard) for a company that supported fair trade for African farmers and opposed sweatshop labour in south-east Asia. Because of the pressure to grow the business and its effect on my own personal time, Mari and I eventually broke up.

Disillusioned with what I considered to be the 'green-lite' world of organics, I quit my job and took time out on a houseboat I had just bought on Bristol Harbour. It was the first time since finishing my degree that I had a chance to think clearly, to reflect and, when no one was around, to cry. I realised I had been keeping myself busy so that I wouldn't have to think about all of the things my culture and I were doing to faraway, out-of-sight out-of-mind places, and the people and creatures that inhabited them. I felt my own complicity in it all, and my own impotence to do anything meaningful about it.

Confused, upset and riddled with Western guilt, I decided to do something completely different, something ridiculous, something that would alter the course of my life in ways I could never have predicted.

~

I'm in the city, to meet an old friend down by the quay, and find myself in a chain bar. The sign on the front door tells me that the toilets are for customer use only, but considering there isn't a tree in the whole city centre under which I could piss without the likelihood of arrest, I decide to honour the spirit of my forebears and completely ignore it.

Above the urinal, no more than a foot from my face, there's an advert. Under a certain kind of logic every blank space is a missed opportunity, and so such innovations are considered astute. This particular advert is for a range of flatulence filtering underwear. I double-check it to make sure it's not a satirical piece of art or subvertisement.

No, it's not, it's a genuine product.

~

Packie tells me that today is the hottest day ever recorded in Ireland. Tomorrow is going to be the hottest day since today. Thirty-three degrees. Knockmoyle has been hotter than Lisbon and Los Angeles over the last week.

I awake naturally, at day-peep, to go fishing on Lough Derg. Pike go deep into cooler waters when things get hot, and so if you want to eat in this weather, necessity dictates you get up early. Ireland is still asleep, and so during the 20 kilometre cycle there I don't encounter a single car. That's fine by me. The sky turns pink, then orange, then blue over the course of the journey, and it is in moments like this I understand the meaning of life.

After a couple of hours, Eugene appears down at the pier, pumping water from the lake into a tank on the back of his tractor. I ask him how he's enjoying the weather. It's killing him, he says. There's been barely a drop of rain for six weeks. No rain means no grass, no grass means no feed for dairy cows, no feed means no money for Eugene. This at a time when small farmers are already struggling with supermarket prices and ruthless competition. The barley growers are in big trouble, too. In Dublin they are issuing fines for anyone caught washing their car or watering their lawns. Drought in Ireland. Strange times indeed. My spring, I'm told, hasn't dried up in living memory. I hope that's not the next record to be broken this year.

I catch one rudd. It's small, smaller than I'd normally kill, but I'm hungry so I pay my respects and eat it raw. I have a wash and swim in the lake, and dry myself by the water's edge with no one else around.

~

On the sawhorse, I saw a log into 60-centimetre lengths using the crosscut. It's a tool quiet enough to fit Wendell Berry's definition of a technology appropriate for the task at hand: 'Will this disturb the sleep of a woman near to giving birth?' As I saw I think of Aldo Leopold's essay 'Good Oak'. In it he narrates the ecological history of Wisconsin, the state where he has a small farm, as he draws his own crosscut through an oak tree felled by a bolt of lightning, the saw 'biting its way, stroke by stroke, decade by decade, into the chronology of a lifetime, written in concentric annual rings'.

His oak had eighty such rings, and was as such only an adolescent. He starts at the year that the 'bolt of lightning put an end to wood-making by this particular oak' around the end of the Second World War, and by the time he gets into the heartwood he is back at the end of the American Civil War. The ecology of Wisconsin, and the US as a whole, changed a lot in the time it took a tree to put on 75 centimetres in diameter.

The spruce in front of me has only twenty-five growth rings, meaning it was planted by a forester (Leopold's oak was almost certainly planted by a squirrel) in 1992, when I was only thirteen, which was also the last time I lived without a mobile phone. In the time it took for this fast-growing tree to fatten up to twenty-five rings, the world became governed by the World Wide Web and countless species had their niches and ancient roles ruled obsolete by mankind. But as the last cut hits the ground, I wonder what stories the rings of trees that I have planted, and watched grow, will one day tell someone else long after I am gone.

~

I walk to Kylebrack to fish for trout. Kylebrack, a neighbour told me, is an anglicisation of its original Irish name, *Choill Bhreac*, which means 'the trout forest'.

There's no longer any real forest left in Kylebrack, and from what I can tell there are no longer any trout in this river either. I try to imagine what this place looked and felt like when it was first

named by its earliest inhabitants, as I put away my rod and walk
home.

~

*Thoreau once wrote that wood warmed him twice: first when he chopped it, and
again when he sat by the fire. Well, either Thoreau was being terse or he had
it a lot easier than me. I've found that wood warms me six times: hauling it
300 metres by shoulder, sawing it, chopping it, stacking it, sitting next to it as
it burns and, finally, by eating the food that it cooks for me.*

*We're prone to forget that, first and foremost, we're heated from the inside.
The Blasket Islanders understood this. Not only did they eat well – very well
by today's standards – their food would have warmed their bones more than
once, which, considering what the weather could be like there at times, was no
harm at all. Sometimes the winds blew so hard the doors would shake and,
when they had to go outside to get fuel for the fire, they would tie themselves
together with a rope to prevent them from being blown off a cliff.*

*Unlike me or Thoreau, the Islanders had no wood available to them for
heating. Not a single tree grew on the island in the nineteenth century,
which is still the case to this day. Like me, they all had turbary rights to
cut turf on the island (I don't exercise this right as bogs are now endangered
landscapes in Ireland, while turf's direct emissions put it in the same
league as coal), which they invariably did using the same kind of two-sided
sleán that is now on display in the Blasket Centre. It was usually the
women who would make the nearly 5-kilometre round trip to collect it with
their donkey, hauling it in wooden baskets known as creels. They supplemented this with gorse and heather, both of which were plentiful and
renewable, along with any driftwood they found which was not considered
useful enough for their homes.*

*To protect themselves against the elements, they usually built their houses
into the hillside. These houses were smaller than most modern living rooms, and*

were built simply from stone and clay mortar from the island. Their doors had latches, but no locks. A dresser and cupboard, across the middle of what modern interior decorators would describe as an open-plan design, divided the house in two. On one side you had the sleeping quarters. Potatoes were stored under their wooden beds. Their mattresses were made from goose down, their bed sheets from used flour sacks, and their blankets from sheep's wool. In between beds they kept a piss pot, which all of the family used throughout the night.

The kitchen side of the house was usually the larger of the two – it was important that it was big enough for dancing, or to set a wake if someone died. Every house would have a sturdy table, built like a kneading trough, with a raised frame around the sides to stop potatoes rolling off and into the dog's eager mouth. Around this they arranged sugawn chairs, made from straw rope – I use one of these myself – while a long wooden couch, set against a wall, was where the night fishermen might have an afternoon nap. It was below this pew-like seat that a family usually kept the hen coop.

The kitchen was the domain of the women, of which three generations would sometimes be found in the same house. In it you would expect to find a skillet, an iron pot, wooden mugs, plates, bowls and a pot oven for baking and roasting on the turf fire. When Tomás Ó Criomhthain was a young man, families would squeeze their cow, donkey and dog into the kitchen at night to keep them warm by the dying embers of the fire. The iron pot was used to soak clothes for washing, yet the Islanders themselves bathed in the cold salt water of the Atlantic. I suppose if you never get accustomed to a hot, soaking, relaxing bath, you're hardly going to miss it.

The Islanders' greatest protection against the elements was their own hardiness, something cultivated from birth. Most men and women, in Peig Sayers' time, would have put on their first pair of boots on their wedding day. Boys wore short trousers, girls home-made dresses. Tough as they were, in both winter and at sea they needed protection from the merciless conditions. The fishermen wore oilskin jackets to keep themselves dry. Their sweaters – the geansaí *– and socks were knitted by the women, who used a spinning wheel* *to turn the wool into yarn. The men took pride in wearing a flat cap, while the women often wore a shawl around their head and neck. The priest always saw their best clothes, whether it be at mass in Dunquin, or in a coffin.*

As I get up from writing, to go outside and pick nettles – Kirsty will later dry the seeds, one or two teaspoonfuls a day of which work wonders for your hair, skin and adrenal glands – I notice it's raining. Putting on my waterproof jacket, which was made in Malaysia, I remind myself that I've a long way to go before I can even dream of calling myself self-reliant.

~

Kirsty has just got back from the city, where she has been busking. She hitched out, something she has been doing more of recently. I admire her courage not to be too swayed by other people's often fearful advice, yet I always find myself glad to see her come through the door safe and sound.

She tells me that she just had a lift with a builder, and that as soon as she got into his van he said, 'You look like a free spirit.' They got to talking. He told her that he used to be too, but that he now needed to bring in €4,500 every week just to pay his lads, whom he feels a lot of responsibility towards. He said the pressure is unforgiving, and that he's not sleeping well any more. Kirsty suggested that he should pack it in, as life is short. Not that easy any more, he said. The bank's involved. Heavily involved. He's bought in and now he can't easily get out.

He asked her what she's up to. Busking on the streets, she said. Can you make a living out of that, he asked? Depends on how much of a living you need to make, she said.

~

I'm meandering through a mosaic of stone-walled fields, trying – failing – to find a pool where two rivers converge, when I meet Michael. He's a farmer, and he owns one of the fields I'm wandering around. It turns out that we met for the first time a few nights earlier, when he was playing the *bodhrán* at the monthly trad session in Holohan's. He tells me that the pub might be closing – young people are leaving rural Ireland, supermarket cans and bottles of wine are so cheap, and Gardaí have been targeting small villages with their breathalysers – and so that session may have been our last.

I've heard better news. That's the second pub, in the space of a week, that I've been told might be closing. Soon the only watering hole around may be our own *síbín*, and the last thing we want is a monopoly. It seems like our free pub is the only one not in danger of going out of business – a peculiar scenario due, perhaps, to the fact it has never been *in* business.

Something needs to be done. But as I stand in a field talking to Michael about hurling, music and the future of farming, I've no idea what that something might be.

~

Casting out into the lough, my mind wanders to past stages of my life, times when I was sure of what I believed, times when *I knew*. Only a decade ago I would have considered myself a vegan and an animal rights activist. Now here I am, actively trying to take the last breath of a breathtaking creature who, if I'm lucky, I'll have to knock over the head before transforming its flesh and brains and heart and eyes into my own.

I cast in again. Still not a bite. The fisherman next to me is throwing in maggots and other bait around his float to attract rudd and roach and perch to his barbed hook. Unlike me, he is taking them out at will, counting them as he goes, before throwing them back in. He's got an awful lot of gear, thousands of pounds' worth perhaps, and 10-kilogramme plastic tubs of bait. He tells me that he just enjoys seeing how many he can catch, and that he lives in hope of a record fish. Says he doesn't like the taste of them anyway. As he pulls in a specimen rudd, he looks very pleased with himself.

I cast in again. My mind has finally stopped wandering, and I focus on the tip of the rod, dragging it every few moments to give the impression that my lure – the head of an old silver spoon with a hook and a streak of red painted onto it – is an injured fish.

The top of my rod bends viciously. I can tell immediately that it's a pike, almost certainly a male jack pike (they are smaller and more likely to go for my spoon bait than the females), but a decent size. He fights like hell. He doesn't want to die – I don't really want to kill him – and I respect his spirit. I take no joy in his struggle, but feel no sentimentality either, as I remind myself why he got himself caught in the first place. He breaks the surface, tries to lose the hook with his ritual death shake, before diving down with all his might. Eventually he tires, and the world as he knows it is about to end. I don't pretend to know what he is experiencing in this moment. Fear? If there is one thing anthropology has taught me, it's that the fear of death isn't universal even among people, let alone species.

I pull him up, and look into his eye as I take his life. I wish him well on his journey, a sentiment I mean sincerely. I'm going to have to face that moment one day too. For a time after, his body pulses. I wonder where his wild spirit has floated off to, or whether I will also take it inside me when I eat his flesh later. The fisherman next to me, who had stopped to watch, tells me it's a fine fish but that it's a shame to have to kill him. I agree.

Cycling back towards Knockmoyle, I stop on an old arced bridge above a shallow, gravelly brook with a pool. I watch a couple of mallards drift over its surface, in no particular hurry. Do they revel in the glory around them? Quietly and unremarkably the female ducks below the surface, before re-emerging shortly afterwards with something dying in her beak. No rod. No rocket stove to cook it on. No ethics. They slowly drift towards the bank, where they come onto land and bask in the final moments of light before nightfall.

Thunder rumbles dramatically, and the rain comes down hard on the way home. I'm wet. I'm tired. I'm alive.

~

It's morning and just before twilight, so at this time of year it's got to be early. Not even the blackbirds are up.

I'm already up in the woods. I need to wheelbarrow around 450 kilogrammes of wood from the woodland to the lean-to, and since much of this has to be done along public roads, this is the safest part of the day to do it. Each barrow-load of roughly 125 kilogrammes has to be pushed for the guts of a kilometre, before returning the same distance for the next lot. That's after hauling each of the logs for roughly 100 metres to the point where they can be wheelbarrowed. It certainly gets the blood circulating before breakfast.

It's at times like these when I wish that my parents had made me push heavy wheelbarrows long distances in the rain when I was a child. That way I probably wouldn't moan and complain about some of the unimportant, stupid stuff I moan and complain about today.

~

I pull the cover off one of the raised beds, whose dry, bare soil has not seen the light of day since it was dug out from the place where the cabin now stands. I'm planting out rainbow chard, winter purslane, endive and rocket which, along with sturdy crops like Brussels sprouts and kale, will keep us in salads for the winter months.

The bed is free of weeds and appears deceptively lifeless, until I spot a furore of frantic activity happening in one of the corners. A colony of ants have had the roof ripped off their home, their world in tatters, their eggs exposed to predators, their future teetering on the edge.

I kneel down above them, and watch intently. At first they appear to be panicked, scurrying around rather aimlessly, yet on closer inspection a pattern emerges from the chaos, with one half finding their way to their eggs and the other going underground through a tunnel against the wooden frame of the bed. One by one they carry their white eggs, which are almost as big as they are, down the hole to some kind of safety. None of them appear to be moaning, none are filming the chaotic scenes for the media or YouTube. They're simply getting on with the job of rebuilding their world, one no longer based on plastic.

A part of me wants to throw the plastic cover back on, but it seems that the more I interfere and disturb the natural order of things, the more havoc I create for others. So I go back to planting out my salads, knowing that one day soon the ants will be back milking aphids and blackfly for honeydew, making people like me question the distinctions between words like 'agriculture' and 'wild'.

~

Watching me writing by candlelight late one evening, a visitor warns me to take care of my eyesight. He tells me that he heard

somewhere once that the master scribes of the ancient manuscripts often went blind from squinting in the dark rooms of
monasteries. There's truth in what he says. He's wearing glasses
himself, the black, thick-framed geeky type, and as he's a young
man I ask him what happened to his own eyesight. He doesn't
know, he says. He just has bad eyesight.

Today is the sixth and final day of his stay. He wanted to get a
break from the city to clear his head, to figure out what he wants
to do with his life, and to see how he finds life in this way. He tells
me that he's enjoyed his week here and that, in many ways, he's
very envious; yet at the same time he's looking forward to getting
back to his creature comforts – his games console, central heating,
television, his music. Especially his music.

Strange, isn't it, he says.

~

After a year of heavy use, my bike needs care and attention. I flip
it upside down and get to work on it. The first job is a fiddly one:
brakes. I stare at them for a short moment. Odd, confusing little
things, I think. On one hand they're going to stop me from careering into an oncoming car and near-certain death; on the other
hand, I'm painfully aware that they're the fruit of a political ideology that is careering head first into a natural world it has forgotten
its dependency on, and near-certain death.

Brakes fixed, I take off one of the wheels to mend a slow punc-
ture I picked up last night. The ingredients of the glue are in a
foreign language, but the three warning symbols are universal:
'Highly flammable', 'Irritant' and 'Dangerous for the
Environment'. I wonder if the danger has already come to pass
before I open it. There's a small piece of PVC-backed sandpaper

and a selection of patches, each one protected on one side by foil and on the other by clear, light plastic. All of this is packed compactly into a green, hard plastic box; which, it seems, is the only green thing about my bike.

I know that none of this is primitive, medieval or even pre-industrial, and it sits uncomfortably on my mind. Such philosophical troubles are clearly a First World problem, yet my dependency on global products like repair kits makes it a Third World problem too. When a man pulls on his brakes in Ireland, whole tracts of ocean and soil are laid to waste in places Western consumers have never even heard of. I tell myself that the bike is a very different proposition to our old Transit van, but only as a matter of degrees, and while it doesn't pump pollutants and emissions into the collective lung, it's still dependent on the same flawed ideology.

The situation, as always, is complicated. In pre-industrial, pre-enclosure times, before people began moving en masse to cities, most of your friends would be living in your parish or village. Nearby streams and rivers were full of fish. But that world went the way of the passenger pigeon, and there's little point pretending otherwise. I don't live in pre-industrial society, the local rivers are dead and my friends are scattered. Yet something inside me still feels that the future – or my future at least – is on foot.

I put the wheel back on, pump the tyres, oil the sprocket and clean out the dirt from the derailleur. It's in good shape. There's a strong argument for not using a bike, but there's no argument for not looking after it well.

I've had this bike a long time, and I've many memories of journeys with it. Once, as I cycled through a wooded area of southeast England called Forest Row, I recall meeting a pure white albino deer – a significant creature in mythology and the only time I've ever encountered one – by the side of the road. I was spellbound, captivated, in awe. Energised by its sight, I was speeding along the road shortly afterwards when a new thought ran through my mind: what would I do if a deer ran out in front of me while I was cycling this fast? I'd seen it happen a couple of times in cars. Moments later a great beast of a stag comes out of the woods and parks himself in exactly the spot I'm careering towards. Wyrd. I slam on the brakes, and for a long moment we stare at each other, before he slowly continues onwards to the other half of the woods.

Crazy as it sounds, I got the sense that this stag stopped me in my tracks to tell me something. I've been trying to heed it ever since.

~

I clumsily spill a wooden tankard of water on the table, and it's inches from soaking my morning's work, which is little more than a sequence of pencilled words on a handful of flimsy, vulnerable sheets of paper.

Warning heeded, I put my work away in a folder, within which is the only copy of my last seven months of writing. It's not saved anywhere on 'the cloud', it hasn't been photocopied, stored on a memory stick or emailed to myself for safekeeping. Its 160 pages would be easily stolen, burnt, soaked or mislaid.

I contemplate copying them out by hand, effectively saving a copy of the manuscript. If I do two thousand words – seven pages – every evening it will take me about four weeks. That's four weeks where I stay up late and get nothing else done. If I don't do it, I could lose seven months' work in a foolish or unfortunate instant.

I put the folder away for now. Que será, será. Trust in the universe. If it is Allah's will. If it is God's will. Fuck it. Or whatever your own way of saying it is. I finish the piece I'm writing, put it away with the rest, and go outside instead. I know that none of it is important anyway, and that the only things that actually need saving are the living, breathing landscapes, creatures and perspectives that rapacious forces want to convert into numbers.

~

Tales of ascending, and descending, with legs.

A couple of years before I quit technology, I found myself at the back of a large crowd of people who had gathered at the foot of an escalator that stretched from the bowels of the London tube system to the expansive universe beyond. I remember thinking someone must have collapsed and that everyone was waiting vigilantly for the paramedics to come. But it soon became clear that the emergency was nothing more serious than a broken-down escalator which had transformed the moving metal steps before the swelling crowd into what was once widely understood to be a set of stairs. If it hadn't been for someone in the middle of the crowd shouting, 'Just use your legs,' there may well have been a crush.

This morning. I've a meeting with someone about this book, and the receptionist of the office complex where she works tells me I need to go to the seventh floor. As I start walking up the stairs, he shouts up to tell me it's the seventh floor, not the second floor, and that the lift is just below me next to the flight of stairs. I tell him that I could do with stretching my legs, and he looks wide-eyed at me, like I'd told him I was about to embark on a solo Arctic expedition. When I get up there the person I think I'm meeting isn't there, as she had no way of contacting me to cancel. I can hardly complain.

Lunchtime. I'm descending a set of stairs, in a train station, on my way to meet a friend. An almost empty escalator rolls on adjacent to me, the exception being a woman and her three children who are static, yet moving downhill at the same pace as me. At the bottom a man – I assume the father – stands facing upwards, with his phone pointed towards his family. What's he doing? I wonder. Of course. He's filming his family as they use an escalator. What else?

~

My taste buds and body have grown to prefer the simple food that I eat, but I sometimes wonder if, when friends and visitors come around for dinner, it looks like I haven't made any effort for them. Ten years ago I would have thought so.

The truth is that it takes a lot of effort. Take this evening's dinner. The plain roasted potatoes (with rosemary) needed weeding, watering and mounding for months, as did the bowls of vegetables and mixed salad. I cycled 40 kilometres and spent three hours trying to catch the pike. Instead, it looks like I knocked it up in ten minutes.

~

On the same evening in 2006 that I decided to sell my houseboat in Bristol Harbour, I climbed out of my berth and stuck a handwritten 'For Sale' sign on its foresail. It was an impulse, but an intuitive one which felt right, and one which I knew I needed to act on immediately, as otherwise I could all too easily convince myself of any of the persuasive, logical reasons why I shouldn't. It almost broke my heart to do it, but the boat was a wonderful part of a life that I no longer really believed in. I was an animal rights activist and environmentalist at the time, yet there I was living a life manufactured from asphalt, plastic, stress, exploitation, oil, shopping centres, rush hours, undrinkable water, polluted air, factory-farmed clothing and synthetic everything. On top of that, I saw that all of my material needs were being mediated by money, which felt like the protagonist in many of our ecological, social and personal crises.

I had come to a point in my life where I wanted to have an economically dependent relationship with the people and landscapes around me, instead of one that was financially dependent on strangers halfway across the world. I wanted to take responsibility for my own material needs and to come face to face with the consequences of my actions. Money, I felt, was hindering me. It enabled me to buy tomatoes from an unknown producer in Italy, soya growth in ex-rainforest in South America, oil from the Middle East and fake leather boots from a factory in China, stuff I didn't need from everywhere, all the while sheltering my senses from the sights, sounds and smells of everything necessary to bring them into existence: oil rigs, quarries, strip mines, the factory system, armies and everything else which I, thinking myself an environmentalist, had been campaigning against. Money enabled me to float around cities, enjoying the fruits of everything I didn't like about the industrial world, without ever having to meet real life – blood, death, shit, dirt – on its own terms. At that, I admit, money excels.

That I needed to radically change my life was clear to me. I wanted to explore what a life without money – a life in direct relationship with what I consumed – might look like. But I didn't have the faintest idea about how one might go about it, or if it was even possible. So I decided to do two things.

The first was to set up a website that would enable anyone, anywhere in the world – me included – to share skills and tools within a neighbourly radius. I used the money I received from the eventual sale of the houseboat to fund the project, which was entirely free to join and had no adverts. Its success surprised me. Within a year it had become the largest skill-sharing platform in the world, with members in over 180 countries, something my little bedroom operation wasn't set up for. Through it people started giving their time and labour for free – no barter, no cash, no credits, no points, no ratings – to help others in their own real-life community with a job that needed doing, often teaching them how to do it themselves in the process. At the time I felt hopeful that we could help reclaim a sense of real community from the jaws of industrial capitalism through the application of cutting-edge technologies.

Seven years later I would come to the conclusion that, if anything, this gift-economy website might actually be making industrial civilisation more resilient, by making it more palatable, a slightly nicer place to be. I wasn't sure I wanted to aid and abet the process of mass urbanisation, drawing people away from the places where gift economies occur naturally without any need for fancy websites. So I decided to merge it with another platform, Streetbank, whose team still felt that complex technology could be a force for good. After putting my heart and soul – and life's savings – into the project for the best part of a decade, it was a hard pill to swallow.

But while that website was still growing in popularity, I also decided that I wanted to walk from Bristol, in the south-west of England, to India without using money. Looking back now I think, 'What a massive hippy.' Needless to say, this turned out to be a monumental failure, mostly down to my own naivety and

inexperience. I made it around the south-west and south coast of
England fine, albeit by losing almost 10 kilogrammes in weight I
didn't really have to lose. By the time I got to France, I was walking up to 80 kilometres per day – sometimes through the night too
– on about half the food I would normally eat while working in an
office. People offered me cash along the way, but under my own
strict rules I couldn't accept it. Things started to go pear-shaped in
France, and after six weeks I returned to England with my tail
between my legs. The media picked up on the story and understandably had a field day. The *Observer* even ran an article, shortly
afterwards, highlighting my incompetence as a way to fail, publicly
and spectacularly, in style.

The public criticism didn't bother me, though I later found out
that it pained my mother, who had been following it all, and that
was the only aspect of it that hurt. That and the fact I felt I hadn't
done justice to an ancient mode of being. But I was as determined
as ever to explore what a life without money, and all that it buys,
might look like. I wanted to test how intimately I could live with
the landscape around me.

There was only one thing to do: try differently.

~

There's good news, and there's bad news. The bad news – for me
– is that sometime over the last few days I've lost a colony of
honeybees. Why or how, I'm not entirely sure. Such things are
unfortunately no longer uncommon – in fact, they have become
frighteningly common – though the reasons differ. A friend lost
eight out of nine colonies to the varroa mite last summer, while
neonicotinoids and mobile phone signals – which continue to play
havoc with their sense of navigation – are said to have reduced

their populations to a fraction of their former size. Human interference, in order to obtain short-term higher honey yields, isn't helping matters either.

The good news – for me – is that they have left a hive full of honeycomb behind, from which we extract honey. Some of this we use for wine (called mead) and beer (later some bottles will be heard exploding in crates), along with beeswax for candles. Not having much of a sweet tooth and needing candles, I'm more interested in the beeswax than the honey.

Last spring we planted red clover on the land around the cabin. Its deep red flowers are out now, carpeting our semi-wild garden with hundreds of red spears among the yellows of ragwort and creeping buttercup, the pink-purples of foxgloves, the white of ox-eyed daisies, the red of poppies and the many shades of green of everything else. As I write on a wooden bench among it all, I notice that the common carder bee and the red-tailed and buff-tailed bumblebees are busy working their way through the red clover, making hay while the sun shines. They've made their own homes in some burrow, and look to be doing well. Roaming among the rest of the garden are spiders, ants, ladybirds, damselflies, dock beetles, wasps, pond skaters and butterflies. Things are as they should be.

I contemplate getting a new colony of honeybees, but decide instead that, for now at least, my time may be better spent creating wildflower meadows. What they need from me, more than anything, is protected habitat. I also decide that whenever I do start keeping bees again, I'll ditch the commercially produced hive that makes interference so convenient and make the kind of skep – a basket hive traditionally made from straw stems twisted and bound together by bramble canes – that broadcaster Alexander Langlands documents in his book *Craft*. An experienced beekeeper, Langlands tells us that out of all his hives, 'I have

absolutely no doubt that the bees in the skep hive fare the best,'
and adds, 'The *cræft* in beekeeping is not in the meddling in the
bees' affairs but in the preparation of their home.'

In this endeavour I'll have to be patient, however. It will take
me a year to grow the correct variety of straw alone. Cræft takes
time.

~

*Having lived in, or visited, many intentional communities – what some might call
communes – in my twenties, I slowly started to build up a picture of why some
places worked well, and why others didn't. As someone who values his freedom
and has heard enough stories of cults-gone-wrong, I've always been cautious of
trying to create brand new communities from scratch, where those involved have
none of the familial bonds or cultural commonalities that are the hallmark of tribal
and indigenous peoples and which, by their nature, can only emerge with time.*

*Some of those I stayed with were fairly dysfunctional, often a raggle-taggle
group of back-to-the-landers, New Agers, lost souls and industrial refugees
who appeared to have few values in common. Even when they did, they often
displayed vastly different levels of commitment to those values. Enthusiastic
meat-eaters would be sitting at the same table as animal rights activists and
vegans, anarchists working alongside those who demonstrably felt that hierar-
chy and strong leaders are needed, the work-shy living with workaholics. There
would be people from England, Spain, Nigeria, Japan, Australia, the US,
China and Argentina all trying to figure out how to live together, despite the fact
that all of them have come to the party with distinct narratives of the world. I
respected their commitment to diversity, and their initial enthusiasm would
often see them through the start-up period. But the differences would inevitably
come out in weekly meetings, and because individuals had no long-standing ties
or connection to each other or to that particular piece of land, they would either
fall away, one by one, or the whole community would simply implode.*

Yet others were working, and working well. From what I could tell these were the places whose inhabitants possessed a sense of common purpose. For some that was veganism and various humanitarian causes; for others, like the Gandhian ashrams or Amish communities, it would be their religious and spiritual beliefs. Personally, I feel incapable of thinking much higher than the ground under my feet, and so on the days when I feel at peace with things I see the world around me as God; the woods, river or mountain as my temple; and my relationship to it all as prayer. On my not-so-at-peace days I find myself just getting on with practicalities, or struggling to overcome one or other of the addictions, habits or expectations I have gained in my thirty-eight years of living in industrial civilisation.

Looking back historically, it's easy to forget that, in the early 1800s, those who left the mainland for the Great Blasket Island did so because they could no longer survive the mainland economy's rents. Because of this, the island effectively became a newly created intentional community of sorts, with a clear common purpose: survival. Not a dog-eat-dog style of survivalism, but one based on values of decency, craft, honour and integrity. Being a small island, its inhabitants were all economically and socially dependent on one another in the realest of terms. They lived and died together. There was no ambulance, no social welfare and little money to bail them out when things went wrong. They had to row their naomhóga *on wild oceans together, and to resist bailiffs together.*

Strengthening those economic bonds was their shared religion – Catholicism – which permeated every aspect of their lives. Their faith in their God, and the Virgin Mary in particular, saw them through many a storm, both meteorological and metaphorical. Peig Sayers would often say that 'God's help is nearer than the door,' despite many personal tragedies, throughout which she would turn her 'thoughts on Mary and the Lord, and on the life of hardship they endured'. There was no church or priest on the island, so they would row into Dunquin for Sunday mass whenever they could; when the Atlantic made that impossible, Peig would say the rosary at her house, and all attended. They celebrated births and weddings together, and grieved the loss of family and friends together.

With the exception of the first generation to move there, most had never lived anywhere else and, like my own father, had no longing to either. They played Gaelic games on the beach. Football was played with a sock filled with grass. Hurling was thrashed out with hurls made of gorse and a ball, called a slio-thar, *made from stocking wool sewn with a hempen thread. They were proud of their cultural heritage, and all spoke in their native Irish, with all of its West Kerry idiosyncrasies. Scholars from the mainland, Britain and Europe would visit them to study Irish, something which played an important role in the renaissance of the language and therefore, it could even be argued, in the political independence of Ireland. Steeped in folklore, many Islanders were fine story-tellers –* seanchaí *– with Peig Sayers being the most notable. They passed many a dark winter night telling stories to each other, or dancing jigs, reels and hornpipes to music composed mostly of the tin whistle, melodeon and fiddle – which some Islanders could make from driftwood washed up on their beach. They sung 'The Faeries' Lament' in the gathering house together, and they gossiped about each other together.*

Left to their own devices I suspect that their grandchildren and great-grand-children might still be there today. But in this global, all-consuming industrial world of ours, nowhere can be left to its own devices, as new expectations and romantic visions of city life spread like wildfire.

And so now the only people keeping the Great Blasket's paths alive are tour-ists, like me, who gawk at the ruined remnants of a people made extinct by the homogenising, all-consuming factories of industrial civilisation. And to think that we call the people who drive such extinctions 'innovators'.

~

After one night of drowning slugs in beer, at the beginning of July, I stopped. It hadn't felt right, and never had. The following night a deer paid our vegetable garden a visit – an unusually brave move considering the garden's close proximity to our cabin – and chewed the tops of our calabrese, sweetcorn, Brussels sprouts, cabbage and purple sprouting broccoli, doing the damage of a thousand slugs in one brief browsing. It was as if Nature had said, 'If you want to do heavy-handed, we can do heavy-handed then.' I decided to listen.

The next day I made a scarecrow out of straw, a few lengths of timber, an old shirt and jeans, a bicycle helmet I no longer used and a Fawkesian mask. It looks so convincing that it still takes me by surprise when I enter the garden. The deer haven't touched the vegetable garden since I put it up.

Watering the plants with the liquid nettle fertiliser I made last month, I notice how strong and healthy the greens look. There is almost no slug damage. Near the long grass and wildflower patch between the garden and the pond I spot three frogs leap out of my way in quick succession, all presumably off to the safety that only wild areas can provide for wildlife. I've not seen that many at one time since I first got here, when the land hadn't been inhabited by people for five years. I notice two are making their way back towards the garden, from which they will eventually have their dinner too and, through doing so, allow me to have mine.

~

Down at the lake. There's a 60-foot barge moored up along the best stretch for fishing, gently rocking in the space between where I usually prey and the spot where the sun meets the horizon at this time of year. I've just cycled 20 kilometres in the hope of catching dinner, and so the sight of this barge is sorely disappointing. Having rejected all of the previous imported sources of protein that sustained me through my vegan years, my only sources of good quality protein now are fish, venison and eggs. For most anglers the barge would be little more than inconvenient, but for me its implications are more serious. I need to eat.

Inside the boat, I can see what appears to be a mother with a brood of five children, all aged somewhere between twelve and twenty. They're all on their phones and tablets – playing, posting, updating, browsing, listening or reading. The mother comes out to talk with me. She tells me that she would love to moor up in this stretch for a few more days – it's picturesque and peaceful – but because there is no electrical hook-up here they'll have to move on, as the children are running out of charge and are on the verge of restlessness. I never thought I would be so glad to hear about young people's addiction to their screens.

The sky has by now exploded into an inflamed orange, the sun reminding us of its controlled prowess before it takes off to wake up some tribe half the world away. I can see a grebe, a cormorant and a heron in different directions, all of whom I suppose have come with similar intentions as me. Out of the boat emerges one of the young women. She asks if I would like a cup of tea or food. I thank her, but tell her that I'm all good for food – I'm not really – as it feels easier than explaining my situation. Shortly afterwards she re-emerges with a cup of tea – milk, half a sugar – a ham and cheese sandwich, a packet of smoky bacon crisps and an apple. It's the sweetest sight you could imagine – kind, thoughtful, everything that speaks well of us as people. To reject such generosity feels wrong, so I slurp down the tea and eat the rest as we chat.

I catch nothing. It's midnight before I get home. I've not had caffeine for over ten years – ham for even longer – and so I don't sleep a wink all night.

~

During my time managing the organic food company in Bristol, I would regularly notice our customers, throughout August and September, walking past brambles heavy with fruit on their way to buy blackberries – £2.50 for a little plastic punnet – from our shop. Even at the time I found that strange. Sometimes, as our customers lurked around our long fridges laden with vegetables and fruit, I would point them towards the bushes outside, and a few would put their punnets back. I probably should have put up a sign saying 'Free blackberries along the path by the car park', but four years of business schooling made that difficult for me to do in those days.

It's early, those moments before the rest of humanity powers up their myriad machines: cars, chainsaws, tractors, strimmers, radios, bulldozers, haulage trucks. I find myself getting up earlier and earlier in search of these moments. I've just been to the woods, twice, to collect wood, two logs each time, one on each shoulder. A practical morning walk. Now I'm picking blackberries for our breakfast. It has been a good year for fruit, so quite quickly I have gathered what I reckon to be about €10 worth of berries; my head, after a lifetime of converting life into numbers, can't help but do the maths. Later we will go picking, as a group, with the intention of making at least 20 litres of wine for our solstice party in the *sibín*.

Seeing the harvest, Packie – with a cheeky glint in his eye – asks Kirsty if she would like to go up the *bóithrín* blackberry picking with him after dark, and she rightly gives him a clip around the ear as he takes off home, laughing.

~

Since the nearby spruce farm was clear-felled in March, a young stag has been coming to graze our grass in broad daylight. He's a refugee, and so I don't know whether to offer him sanctuary and protection, or to do what I would do if we were living in peace with his kind: which would be to kill him, skin him, butcher him, smoke him and use every sinew and bone in his body. I watch him now as he slowly and gracefully makes his way through our copse, his impressive antlers quietly moving among the long grass, willow, sweet chestnut and hazel, grazing on all as he goes.

I weigh it up. The old animal rights activist in me says no, that my way of life has brutalised his long enough, and to offer him and his tribe a place of refuge. The conservationist in me is confused; there's more deer around here than there is habitat for them now. That's not because there are too many deer, but because there isn't enough habitat. The primal hunter-gatherer in me says I should kill him, and to view the act as part of the only culture that has ever made any sense to me.

As I watch him, I ponder. I remember Aldo Leopold's words, penned on his own smallholding in Wisconsin, from his short essay 'Thinking Like a Mountain'. When Leopold was young and full of trigger-itch, he 'never heard of passing up a chance to kill a wolf'. At the time he believed that 'fewer wolves meant more deer' and therefore 'no wolves would mean hunters' paradise.' But one afternoon, after he and his friends had pumped lead into a pack of grown wolf pups playing with their mother, he witnessed for the first time the 'fierce green fire dying in her eyes', and came to 'suspect that just as a deer herd lives in mortal fear of its wolves, so does a mountain live in mortal fear of its deer'. A mountain without wolves, he explains, looks 'as if someone had given God a new pruning shears, and forbidden him all other exercise'.

I know this young copse lives in mortal fear of this stag, and his kind, entering its boundaries. And then I wonder if I do in fact know, or if there is something that I do not understand about this hill and its creatures yet. Unsure, I decide to leave things as they are for now, and to wait for the answer to come.

~

After nine months of splashing warm – and sometimes cold – water over myself in the Victorianesque aluminium bath tub that usually hangs on the timber-frame outside, I feel motivated to make an outdoor hot tub. September is a good time of year for such jobs, as I've more spare time and there's usually plenty of dry weather.

My friend Matt has a garden full of salvaged cast iron hot tubs, and he kindly promises to bring one over the next time he's visiting. The most important part of the job is picking the perfect spot for it to go. There are many factors to consider. Ideally it would catch the afternoon and evening sun, so as to gently warm up a bathful of cold water before the fire is lit below it, thus saving wood. Privacy is important, more for my neighbours' sake than our own, as I'm not sure Packie, Kathleen or Tommy are quite ready to see me climb out of a hot tub naked. I want the tub to be surrounded by trees – to protect its cob walls from the elements and for the calming lack of urgency that they often inspire – while at the same time having unobstructed views of the Milky Way, for those times when I'll want to bathe at night. And it needs to be somewhere the water can freely drain away without creating a mud bath.

After a week of mulling it over, I decide upon a site near the centre of a triangle of which our cabin, The Happy Pig and the farmhouse are the corner points. I level off the ground, which is next to the fire-hut, and put down stone foundations on which the tub will eventually sit. I dig a small French drain, starting below where the plughole will be and running all the way into a bigger, older drain. With all the groundwork done, Matt and I lift the tub into place. Being built to last many lifetimes, it's heavy, but it sits perfectly in its place, level except for an imperceptible drop to the plughole end, to allow the bath to fully drain after every use.

Running along one length of the bath I make a bench, where glasses of blackcurrant wine and clothes can rest clean and dry. Around this I build a rockery from more of Tommy's stones, on top of which cob will graduate into the top lip of the tub. Cob – a mixture of clay, straw, sand and water – is a wonderful, forgiving material to use, but after mixing over 30 tonnes of it by foot (you stamp the ingredients into cob in much the same way you traditionally make wine out of grapes) over the course of two summers when building the hostel, I still feel sick at the sight of it. We have lots of good quality clay, however, so all temporary qualms I have about using it are set aside, and it's what I use to decorate and insulate the sides.

Along the top of the cob I embed a selection of coloured tiles which I found when I first moved here, which will offer it ample protection from splashing water. Both the firebox below the bath, and the chimney that emerges out of it, are also made out of cob, and they take me most of the day to get right. As the light begins to fade and the midges briefly turn heaven into hell, I smooth off the cob surrounds with my fingers to give it a natural, textured finish. Standing back, looking at the day's work, it's a pleasing result. Some people create art on the cob – sculptures of the sun, flowers, dolphins and suchlike – but I haven't got an artistic bone in my body, so I keep it plain. The whole thing costs me €15, which I paid to the local waste recycling plant for a couple of lengths of insulated flue pipe.

It will be a few days of dry weather before I'll test it out. If I've forgotten something critical – it's my first hot tub, so it's a definite possibility – I may have to rip it all out and start again. Right now I'm covered from head to toe in sweat and cob, so I take down the aluminium bath and pour cold water over my warm body while the new hot tub teases me in the background.

~

Today's lunch: a bowl of salad – mustard lettuce, rocket, rainbow chard, horsetail, calabrese leaves, fennel, spinach, peas, parsley and grated courgette – along with boiled eggs and a fresh mackerel which, though gutted, is otherwise raw and whole.

I wash it all down with a small bowl of mackerel blood. For those unused to drinking blood it can taste strong and intense, but from the moment it enters your body it is difficult to imagine a more potent drink on the planet.

~

I wake up suddenly. I've no idea what clock-time it is, but the dense darkness cloaking the cabin gives me the impression that it's the dead of night. I realise that Kirsty isn't beside me. She said she would be home early tonight, and so all sorts of scenarios start running through my mind. I've no way of contacting her, and she has no way of contacting anyone. We always tell each other not to worry if one of us is out much later than expected, but it's easier said than done when it comes to people you love. I try to go back to sleep, remembering that it's not so long ago that not knowing everyone's every movement was normal. I tell myself that this is Knockmoyle, not New York, and that if she has been in an accident then enough people here know her and would undoubtedly get word to me immediately.

The darkness is fading as I get up. It's very late. Or very early. I couldn't sleep. I'm on my way out the door to gather firewood when the latch goes and in walks Kirsty, looking very pleased with herself. She's had a terrific, spontaneous night of dancing with a few girlfriends. I hold her tight and tell her that I was worried. She smiles and tells me to either get a phone like the rest of the world – there are now more phones than people on the planet – or to make peace with just not knowing sometimes.

~

Out blackberry picking with Jorne (or, as the neighbours call him, Captain John), who lives in the farmhouse. This, he tells me, is now considered poaching in his homeland, the Netherlands (from which he considers himself to be an industrial refugee), and is punishable with a hefty fine.

As we continue ambling up the verges of the *bóithrín*, we wonder at what point in Dutch history such an imposition became acceptable, and I think about what I would do – above and beyond simply ignoring it – if such legislation were to be passed in Ireland.

The next morning, on an adjacent stretch of road, a man in a high-visibility bomber jacket and trousers races up behind me on a quad bike, and into his two-way radio I hear him say, 'It's a man, I think he's just picking blackberries.' He asks me if I'm going to make jam, and tells me to watch out for the three tractors coming up behind. It would be hard not to. They're huge, each with hedge-cutters attached, doing the type of rough job which the human hand of a hedgelayer would find unbearable.

I ask him if they'll leave the patch I'm picking from alone. He agrees, and I thank him. The tractors pass me by, each with a wave from the drivers, and make their way down the road. For the next 20 kilometres not another inch of hedgerow will be spared. Blackberry season, on this road, is over just as it started.

I walk back home, and have enough for breakfast.

~

I awoke to a flurry of phone calls. It was the morning before Buy Nothing Day 2008, the day on which I was due to quit using money, and the media had got wind of my plans. Both my mobile phone and the landline of the house I was about to move out of were ringing at once, and continued to all day; by nightfall I must have given close to forty interviews to press and radio around the world, all asking more or less the same questions. The world's financial economy had, coincidentally, just gone into meltdown, so for the first time in a generation people were seriously wondering how they could live on less – much less, in some cases. There was unprecedented anger towards the banks and big business, and there was a deep questioning of the global financial system that hadn't happened before in my lifetime. At the time, I was twenty-eight and impassioned, so I decided to take the unexpected opportunity to speak out about my reasons for embarking on what was turning out to be quite a timely endeavour.

It was strange doing all that talking about it before I had even begun living without money. All I had really wanted to do was explore what a life without money might look and feel like, to see if it was even practically possible in the modern age and, if it was, what the practicalities were. After all, there was little point in talking about the intended and unintended consequences of money – on our societies, landscapes, cultures, economies, ecologies, our physical, emotional and mental health, our spirituality – if such a life were to turn out to be a form of living hell. I had wanted to give my mouth a rest in order to give my hands a chance, and so, while I played the hand I was dealt, it wasn't the ideal start to what was going to be an entirely new way of living for me.

I had more important, immediate things on my mind too. I had already committed to putting on a three-course meal, made entirely for free out of foraged and waste food, for 150 people the following day, which would also be my first without money. As I grew tired of the sound of my own voice, I was acutely aware that I still had only a fraction of the food needed for that many meals, and that I was about to start something – originally intended to be for a year – that I had no idea if I could finish. To add to the pressure, the eyes of the world's media were back on me again.

The next day came and went. Before I knew it, it was the final day of my year, and sixty volunteers and I were putting on a follow-up feast; except this time it was for over a thousand people as part of a day-long free festival of music, cinema, talks, workshops and free shops. In between those two feasts had been the most deeply affecting period of my life, the adventures of which are the subject of my first book, *The Moneyless Man*. I had achieved what I had set out to achieve, but it no longer felt like an achievement. Living without money had become as second nature to me as living with money had been before; or, perhaps I had finally tapped into my primal, first nature.

By that stage I had already decided to continue that way of life for as long as it felt right. The thought of using money once again to mediate my relationships began to feel absurd, unnatural and undesirable. I had stopped missing things and had come to love what I had gained. Why give that up?

After three years, I decided to start using money again, for a time at least. There were a few reasons. First, the whole undertaking was starting to consume me. The media's fascination with it was one thing – I could have, theoretically at least, quit all of that and just simply lived the life, without being its spokesperson – but it was all that anyone I met would ever talk to me about. Strangers would even pull me over in the street to give me their opinion.

I had also come to the point where I wanted to set up a land-based community where others could come and experience some time – a day, a weekend, six months, a lifetime – living in a direct, immediate relationship with a place. Along with that I realised that the only thing I was really missing was my family, who were in the same place as they had always been. I had been living in the UK, at that point, for over a decade, and I was lucky if I saw them once or twice a year. Like myself, they weren't getting any younger, and I wanted to spend time with them before it was too late. So I decided to move back to Ireland, and set up a small community there, using the proceeds from *The Moneyless Man*, which had already been translated into twenty languages. I wanted to use these unexpected funds to create a place where others could also live, rent- and mortgage-free, in a direct relationship with a landscape. The real challenge, I now know, was only beginning.

I went to a pub with some friends in Bristol and handed the barman a piece of paper in return for ale. It all felt surreal. Stopping being moneyless felt even more strange than starting. I knew that I wanted to get back to living directly from my landscape again, and I knew that I wanted to do it at a more primal level, with none of the distractions that I had found ways of accessing, even without money. To do so I knew that, one day, I was going to have to relinquish all of the things that were mediating my relationship to the land and preventing me from cultivating an intimate relationship with everything in my immediate surroundings. It was a hard, and yet exciting, thought.

We finished our pints, and a few more, and left. Some of the people I went out with that night I have never seen since.

~

People often accuse me of being a Luddite, a term that wrongly gets used synonymously with words like 'technophobe' or 'antiprogressive'. I respond by saying that I'm not worthy of such high praise, but thank them nonetheless.

As I sit in a cabin writing, I recall how, between 1811 and 1814, the actual Luddites rebelled against the wealthy industrialists and their powerful political friends who, at the time of the land enclosures, were obliterating the cottage economies which afforded the common people a familial, purposeful and pleasant life. By then the steam engine – what Carlyle called 'Stygian forges with their fire-throats and never-resting sledge-hammers' – was enabling one man to do the work it would have taken two or three hundred men to do only a decade earlier. This effectively reduced a once proud and independent class of skilled craftspeople who had worked in their rural cottages to, at best, wage slaves in an urban slum and, at worst, the unemployed in an urban slum. After a three-year resistance the industrialists eventually won, and the rest, as they say, is history.

As I sit in a cabin writing about the Luddites, I fantasise about rebellion. But where to start? I look around and see that, in the twenty-first century, the machine is everywhere, even in my own head. Maybe that's not such a bad place to start then.

~

Buying books, during that twenty-year period of my life in which
the internet pervaded much of what I did, was quick and simple.
Go online, search for the book you want (they have it, always),
click a few buttons and it's in your letterbox within a few days. If
you're not the disciplined sort, you may even get a couple of other
books in the post with it. These go on the bookshelf as intellectual
wallpaper, and you reassure yourself you'll make time to read
them soon.

Easy. Probably too easy.

Not so easy any more. I'm looking for a specific book, *A Social
History of Ancient Ireland*, published over one hundred years ago but
out of print now. I came across a free online version of it a number
of years ago, but wasn't motivated enough to read it at the time.
In it, the author describes how my ancestors lived pre-industrial-
ism, including everything from how they grew food to the details
of their legal system, Brehon Law.

With my thumb out, I stroll down the road not long after
sunrise, and make it to Charlie Byrne's bookstore in Galway City
just as its doors are opening. It's an independent shop, selling
mostly second-hand and antiquarian books that are crammed into
tall shelves in its many nooks and crannies. Vinny, whom I've
come to know a little over the last few years, is working. He seems
to know every one of the many thousands of books in the shop,
and exactly where it might be.

I ask him for the book I'm looking for. He hasn't got it – it's
rare. Very rare. It will be a week or more before he can find a copy
of it. He tells me about an event, happening the following week,
with the writer who most influenced my decision to stop being so
dictated by clock-time. On my way out the door I notice a book
on the shelf that catches my eye and calls me. It's the most beautiful book, as artefact, that I think I've ever seen. It's called *Nature:*

or, The Poetry of Earth and Sea by Madame Michelet. Its hardback cover is embossed with an illustration which is hand-crafted and gilded, as are the edges of its pages. It closes with a good, healthy thump. On the back and inside covers there are no words of praise from the media or notable persons, only a hand-written message saying 'From Grandpa & Grandma, 1880', written in the same year the book was published. It contains two hundred illustrations by the artist Giacomelli (who also illustrated *The Bird* by Madame Michelet's husband, Jules), a feat which took two years alone.

I learn more from simply looking at this book – as artefact, as art, as craft – than I have done from the contents of most others. It costs an arm and a leg – as it should – but I console myself with the thought that I'd rather have only ten of such books in the cabin than an entire library of cheap paperbacks on nature and craftsmanship whose production betrayed their content.

I say goodbye to Vinny, tell him I'll see him after the event next week, and start out home. I make it back by lunch.

~

Holohan's pub closed its doors a few days ago. Ten years ago there were two pubs in the village of Abbey. Now there are none.

I meet a local in the next-door shop. Awful shame, he says. Where do we go for a pint now? Nowhere, I say.

~

I've never tickled trout before. While tickling is still commonly used to catch fish in the Falkland Islands, it has been banned almost everywhere else. Why? I'm not sure. To my eyes, it seems to be the least cruel method of fishing, and it is certainly the most primitive, requiring not a single piece of gear or bait; slowness, awareness and sensitivity in the touch, along with a good understanding of the trout's nature and habits, are just about all you need. Some claim it's unsporting, but I've no interest in gentlemen's rules. I don't fish for sport, I fish to eat, as outdated as that might sound. Yet even the sporting argument doesn't stand up to scrutiny, as tickling requires greater skill and knowledge than simply casting a €4.99 spinner, with an acid-sharpened hook on a reel of monofilament line, into a lake.

Because of tickling's illegality, my friend is obviously only teaching me the theory of how a Falkland Islander, or poacher, would go about it today. We wouldn't dream of catching a trout, killing it and having it for dinner. Waist deep in water, we move slowly, unconcerned about making noise or scaring the fish away. We want them to be scared, and to swim away from open waters into the apparent security of the banks. My friend tells me that good knowledge of a river, built up over years, is at the heart of good tickling. He leads me on towards an overhanging tree, the kind of place an alerted trout would take refuge; if I were being hunted I'd make it as difficult as possible for my predator too. The problem in the modern world is that it's hard to know who our real predators are anymore.

We run our hands nervously – there are, allegedly, crayfish in this river too – along the bank, searching out the kind of holes rats make and brown trout like to hide in. A poacher would now try to feel the smooth, fleshy, alive belly of a trout protruding out of one of these holes. He would aim not to flinch as he felt it, but instead would slowly and gently run his hand along the underbelly of the fish, tickling it, before moving his other hand in for the kill. Once the fish is relaxed and acquainted with the gentle stroke of the hand, the poacher would grasp and bend the fish with a

lightning-quick motion, pull it into his stomach and cast it onto
the bank, where it would be knocked over the head as quickly and
painlessly as possible.

Not being poachers, we leave empty-handed, vowing to return
one day to do some real fishing with a kitbag of industrial gear.
We dry ourselves off as we walk home in the late evening sun.

~

The lunar landscape before me is timeless, spectacular and unique.
I'm with a friend on top of a hill in the Burren National Park, a
1,030 square kilometre World Heritage Site and national park in
County Clare, the edge of which is a four hour cycle ride from our
smallholding.

It's renowned worldwide for its wildflowers, which – like the
oak – appear to prefer tough terrain to tough competition. I'm
surrounded by pinpricks of yellows and purples and oranges and
blues and reds, coming out of the cracks in the limestone, like a
rainbow squeezing its way out of the moon. Rumour has it that
J.R.R. Tolkien, who spent plenty of time here, based his map of
Middle-earth on the Burren and, if you compare both maps, there
appears to be some evidence to back up the claim. Considering
the financial rewards of being the real-life Middle-earth, however,
I've no doubt that there are many regional tourist boards making
claims to all things Tolkien. Money and truth seldom go hand in
hand.

To our west-northwest, far off in the distance, is the fishing
village of Kinvarra, where the sun is making its way under the
horizon, turning the world beyond it an intense pink. Not long
after, a full moon is rising golden to our east-southeast, drawing
the tide gradually closer to all that dwells in the fishing villages

below us. Between us and the Atlantic Ocean there's a patchwork blanket of fields stitched together by gigantic Burren stone walls. My back feels the slightest breath of a breeze, like a satisfied sigh at the end of a good day's work, and it's welcome. There is barely a single perceptible sound other than our own footsteps, and as that becomes noticeable we sit down together on a smooth hump of this karstic terrain.

I feel like I want to get down on my hands and knees and worship what is before me. After a short moment sitting, my friend is on her feet again, filming its glory with her smartphone, and posting to her social media accounts.

Quite astonished, she tells me that she has just noticed that you can now buy 'likes' for the things you post. I suppose, under its own logic, that makes sense. Money has always been able to buy a fake sense of popularity in the real world too. It was only a matter of time before this principle would be applied to the virtual world.

The sky ablaze with indifference, we descend the hill before the light fades out completely.

~

In among the rubble of ripped-up earth, chewed-up spruce and spat-out brush, a purple blanket of foxgloves has suddenly risen up, alive and defiant and spectacular, in the place near our small-holding which is now called a clear-fell. There must be thousands. There may be hundreds of thousands.

It's a beautiful, miraculous sight. It's enough to give a man hope.

Autumn

We all strive for safety, prosperity, comfort, long life, and dull-
ness ... A measure of success in this is all well enough, and
perhaps is a requisite to objective thinking, but too much safety
seems to yield only danger in the long run. Perhaps this is behind
Thoreau's dictum: In wildness is the salvation of the world.

Aldo Leopold, *A Sand County Almanac* (1949)

September equinox. Another season. We're having a fire gathering – or, as it's traditionally known, a massive piss-up – this evening, and it has come to my attention that I need to knock up some
outdoor seating quite quickly. We've got a free and easy supply of
discarded pallets, and considering there's almost nothing that you
can't do with a pallet, the stack behind the wood store is central to
my plans.

Making a pallet chair is child's play. I saw a pallet in half at a
slight angle, so that one half (the back) reclines on top of other half
(the base). I attach a couple of spruce boards as supporting arms,
before raising the entire two-person seat up on another pallet.
That's one done, and I've barely got going. It's the type of chair
that would be called trendy and rustic in London, and for which
time-poor, money-rich people would pay a silly price.

Job done, I decide to take the afternoon off and to spend it
reading in the sun. It is in these moments I find myself having to

battle the work ethic with which I've struggled all my life, the one that has always told me that such afternoons are unproductive. As the sun descends behind the ash and horse chestnut trees to the west and people begin to arrive, I reflect on the fact that I've spent much of my life producing little more than bullshit anyway.

~

Setting the fire, part II.

I'm using an old newspaper, which I found in a neighbour's recycling bin, whose headlines I conclude have probably already been recycled enough across social media. As I sort the tinder layer, I open a supplement on technology. I normally ignore the print and try to see the newspaper for what it actually is – thin slivers of wood pulp that are excellent for lighting – but there's an oversized photograph of a sex robot staring at me on the front cover. As intended, it has my attention.

The supplement claims that, within the next ten years, the sex robot industry will be big business. It certainly will if the editors of international media corporations say it will. This piece is spotlighting all sorts of products. There's a virtual reality machine which runs a sort of interactive porn that connects with 'a sleeve' that attaches to the penis and, I can only imagine, replicates whatever action is taking place on the headset. You can pick any girl you like – blonde, brunette, buxom, dominant, passive, young or old. A different girl every night, if you like. None will refuse you because you were being an inconsiderate idiot all day, and you'll no longer have to deal with any of the messiness that comes with human relationships.

No one is to be left out, or perhaps spared. There are plenty of options for women and the LGBTQ community too. You'll also be able to buy an actual robot you can go to bed with, and they'll tell you all the dirty/romantic/caring/poetic/raunchy/sweet/sadistic things you like to hear, depending on what preferences you select from his/her/its drop-down menu (at some point in the future someone will get sued for calling a robot 'it'). There's an interview with one woman who has fallen in love with her sex robot, and says they plan to marry. She's not alone.

Later I speak to a friend about it. He tells me not to worry, it'll never get to the point where sex robots replace the need for actual intimate, human sexual relationships. He might be right, as it's too early to tell. Either way, will it mean that sex with living, breathing women who don't like deep-throating, or with men who haven't got six-packs and vibrating penises, will seem mundane, unexciting and undesirable in comparison?

One week later. A farmer from across the way comes up the *bóithrín* in his old red Massey Ferguson. He stops to chat for a moment, before continuing on to one of his fields, where he has to drop off a big blue drum of water for his horses. Shortly afterwards I see him back again, this time with a large round bale of silage, which he promptly puts next to the drum. I notice how, in the entire process of looking after his animals, his feet don't touch the ground once. Rather remarkable, really.

I'm sure if someone told the farmers of yesteryear that there would come a time when a farmer's feet wouldn't touch the living, breathing earth under them, they probably would have said that it would never come to that, either.

~

October is a good time for small, important jobs, like sharpening your tools. Even though most won't get used again until the spring,

I like to know that they're ready and capable of doing a neat, tidy job in advance of when they are needed.

I start with the scythe. As it has become blunt through the endless cycle of use and sharpening, it first needs to be drawn out with a peening jig, a process that involves cold hammering the edge of the blade to give it a good cutting profile. Once it has lost a bit of its flab, it's ready for sharpening. There are two tricks to sharpening a scythe safely. One is to keep the length of the blade secured firmly against your arm; the other is concentration. Lose concentration and expect blood. Meditative states, I've found, are usually easier to achieve when such existential matters are involved. A well-sharpened scythe is the difference between back-breaking hardship and a pleasant afternoon; a field of semi-flattened grass, or rows of good hay. The most common mistake people make when scything is that, in an attempt to save time, they don't sharpen their tool as soon as it has lost its keenness, and thus spend the time tutting and cursing instead of enjoying a satisfying experience.

The double-handed crosscut saw is up next. Honing each of its teeth with a file can take you half the morning, but it will pay you back with interest later. The rest of the tools – the chisels, machete, clippers, pruners, sickle, drawknife, axe and knives – I work with a hand-turned, rotating whetstone mounted on a workbench. This requires more concentration, as it is much easier to lose your edge than it is to sharpen it.

~

I receive a letter from my literary agent (most publishers won't even read your manuscript if it doesn't come from one). She tells me that the *Guardian* articles I have been writing over the course of the year have caused a stir, and that's she's getting a lot of requests for TV and radio interviews regarding my reasons for unplugging and the practicalities of doing so. They all want to run interviews that afternoon or evening, or possibly that week, but none are interested in waiting for a month, which is about the length of time it would take us to organise it all via post. I once understood the rush to do these things, but I don't anymore. She says we're losing lots of opportunities to raise my profile and increase the potential for a good book deal (which, after almost a year of writing, I still don't have). I'm sure she's right.

I write back. So be it.

~

It's the first time I've picked up the tin whistle in twenty-five years. The woman who last taught me is long dead, as is my ability to read music. Though now that I think back, I'm not sure if I ever could. They say that if you can play music by ear it is very difficult to learn how to play by reading it, and vice versa.

I'm sitting in The Hill. Once a week a group of local musicians gather to practise for their live session on the second Saturday of every month. But, deep down, music has always been an excuse for a social. Feeling unworthy of playing with the folk gathered around the table – they're all accomplished musicians – I'm sitting at the bar, listening in, trying to get an ear for the jigs and reels they're playing, and otherwise keeping my head down. The tunes are intricate. Too intricate for me.

One of the musicians, Ned, asks me if I have an instrument with me. I pull the tin whistle from my back pocket, where I had concealed it. Take a seat, he says. I tell them that I don't want to disrupt their session, or their routine, but the flute player Mike

– a man renowned in these parts – asks me to play any song I know. Before I know it, each one of them – with their fiddles, squeezeboxes, *bodhráns*, flutes, tin whistles, mandolins and banjos – is playing along with me. I'm entirely out of my depth, but I'm not made to feel it. As each woman and man's genius weaves itself together, I feel part of something bigger and more important than myself.

Which is all I ever really want to feel.

~

I come upon a deer on the side of the road. This is a more frequent sight at this time of year. As deer are rhythmic creatures and more active around dawn and dusk – exactly when commuters are busiest toing and froing around the start of November – the clocks going back means they cross roads between woods and pasture at levels of daylight during which they would have been safer staying put.

The doe's guts have been spilt, and a crow appears to have taken out one of her eyes. Nutritious. But it is a cold, crisp day and there are no bluebottles near her yet. I put my hand inside her body, and find her still warm. She should be good to use. The problem is, I've not butchered a whole deer before. I had been waiting for the deer season to reopen for six months, but as I found myself busy with other things, this has caught me unprepared. In some ways it is perfect, as it means I don't have to kill a deer, the thought of which my civilised mind has never found easy. But in other ways I feel stupidly rushed. I had intended to read books about how to butcher deer, and make use of every part, well in advance of the event. Now I have a matter of hours to make decisions which I feel a lack of competency to make.

Back at the smallholding. I hang the deer, by her neck, off the skinning frame in the fire-hut. I sever and snap her front legs, which are single jointed, and make incisions into the skin, both around the neck and where the brown and white hair meet on her legs. This much I do know how to do. Being fresh, the skin peels off relatively easily, leaving a red, raw, lean and muscular body dangling on a rope. Only the head now resembles what I once considered to be a deer, and her intact eye looks right at me.

The light is already fading. Just as my gardening books are soiled with muck, my butchering book is now splattered with blood. I realise that I should have left the skin on, for a day at least. Mistake number one. Nothing to be done but keep going, as I can hardly put it back on now. It's getting difficult to read the book, so I make cuts instinctively. Slight mistakes two, three, four and five, but nothing fatal, as it's all just meat and tallow and it's not as if I am selling to the public. With all of the meat cleaned from the bones, I saw off the poor beast's head, from which I'll later finger out the brains to use for tanning the skin. I put the meat in containers, along with her heart, liver, lungs and sinews, and take it away to hang for a few days, where I will later make mistake number six. It's now dark, and I'm tired, and in need of food.

I feel disappointed with myself. If I had prepared myself properly I'd have done a much better job. But I didn't grow up like Huckleberry Finn or Tom Sawyer, and there have been no elders in my life to link the ancient with the present. The maps of my journey home are scattered and piecemeal. As I attempt to bring these maps together under one whole, I have to accept that wrong turns are inevitable along the way.

Changing out of my clothes in the cabin, I become aware that I've been wearing a yellow T-shirt I bought from an animal rights charity in Bristol ten years ago. 'Stop animal testing,' it reads. It's a sentiment I still strongly agree with – there's never an excuse for unnecessary cruelty – but right now it's covered in deer and sheep blood (a neighbour had dropped over a fresh sheepskin that needed fleshing the day before). When I bought that T-shirt I couldn't have imagined my life as it is now. I thought that people like me were big-time idiots lacking empathy for other sentient beings. But back then I lived in the city, and I lived a city life where I was insulated from the violence and cruelty my city ways were dependent upon.

Recalling some of my old animal rights comrades, I've no doubt some would disown me now. To them my words will probably feel treacherous. It's a thought which saddens me, as having held their views for long enough myself, I understand and respect where they are coming from. Yet I have to remind myself that it wasn't me who killed this deer. A car did. Cars aren't vegan. Phones aren't vegan. Plastic tubs of vitamins aren't vegan. Chickpeas, soya and hemp seeds – none of it is vegan, not really. It's all the harvest of a political ideology which is causing the sixth mass extinction of species, one which is wiping out one habitat after the next, polluting rivers, soil, oceans and every breath of atmosphere as it spreads. From where I stand now, this bloody carcass in front of me feels more vegan than the plastic packet of cacao nibs I once sold, or the T-shirt – dyed yellow with who knows what – on my back. Well, if not more vegan, at least more honest.

Looking at it in my hand, I notice how similar her heart is to my own. How many people passed her on the side of the road today? I pay my respects to the life it once was, store it in a cool place, and leave the rest of the work until morning.

~

The clocks going back doesn't only affect deer. For almost every-one – excepting babies, the comatose, myself, rocks, fish, trees and wildlife – it meant an extra hour in bed this morning, followed by six months of seemingly darker evenings. People will, for weeks to come, inevitably comment on the terrible, sudden shortness of the days, but today or tomorrow won't be getting shorter any quicker than any other day since summer solstice. Untied from clock-time, this evening won't feel significantly shorter in length from yesterday.

Still, the days certainly aren't getting any longer for a time, so I ought to stock up on candles. Having fields dotted with rush, I have an endless source of candle wick. I cut a clump of its green, cylindrical shoots, and peel them one by one. The outer skin comes off easily enough, leaving a soft, dry, absorbent pith, which is what I am after. I fire up the rocket stove, cook dinner and, once it's done, melt a bowl of beeswax in a pot of hot water. It smells as good as honey straight from the comb. Taking care to keep an 8 centimetre length of rush pith centred, I pour the melted beeswax around the wick and into one of a collection of shot glasses I once found in a charity shop. I hold the wick steady, keeping it in the middle of the glass and touching the base – as I watch the beeswax gradually solidify again in front of my eyes.

Making a candle is easy. The real craft lies in the first part of the process: the keeping of the bees. Actually, the most difficult part of candle-making is deciding to reject electrical lighting.

~

I'm on a rare foray into Galway City, where I'm due to give a talk to a 'Slow at Work' group about my experiences of living without technology. On my way in, I'm struck by how many more people are living on the streets, homeless, than when I lived there as a student.

I get to talking with one man who is sitting on the ground with his dog. I ask him how life is. He says it's hard, and that he feels shit about begging. He never thought he would be a beggar.

I tell him we're all begging – pushing our wares and services on social media, singing our own praises, trying to convince people that they really need what we have to offer – and that he's probably the only honest one among us.

You're right, he says, and laughs, and we laugh a while longer before I head off towards a posh venue where I will also be staying for the night.

~

The week before I quit email, phones and all electronic modes of communication, I sent a group email to the thousands of contacts I had accumulated over twenty years of life spent trying to embrace the new ways. The message simply informed people that I would no longer be available by email, that I was leaving Cyberia, while letting them know my postal address, something only a handful of them would have known. It went to a mix of close friends, family, ex-colleagues, ex-girlfriends, collaborators, acquaintances and random strangers who, at one point or another, must have got in touch about something neither of us can probably remember.

The reaction, initially, was overwhelming. Most days, in the weeks that followed, I would find my postbox stuffed. I imagined the postman wondering what was going on, as all I had received from him in the past were bills, or official letters with clear plastic windows and automated addresses. Now addresses were hand-written, often in colourful or hand-crafted envelopes. Some were from close friends, wondering how we were going to see as much of each other from now on, while others were from names I didn't recall, wishing me all the best or explaining to me why I should minimise my use of electronics rather than rejecting them outright. On top of that I was receiving letters, via the editor at the newspaper, from readers who wanted to share their opinion about my decision to unplug.

It's now a Friday in October, almost ten months later. I check my letterbox for the last time this week. Unlike email, it's pointless checking it more than once after the postman delivers, or over the weekend when he doesn't come at all.

There's a letter from my mother, telling me that an old neighbour has died. It's too late to go to his funeral, as people are buried within three days in Ireland. Other than that, nothing. No letters from acquaintances, no junk mail, no random strangers, no close friends, no bills. Whereas in January I was spending as much on stamps as I had previously done on a mobile phone, now I'm only spending a fraction of it. It's an odd mix of feeling forgotten on one hand and, on the other, feeling liberated from relentless communication with people who, in all likelihood, live too far away for our relationship to deepen.

Those letters I do receive tend to be from people who really want to get in touch. Inconvenience is a great filter.

~

It's bin collection day. Out here it comes once every two weeks. We don't have a bin, but everyone else puts their bins out as usual. Each time, before the waste company comes, I walk down to one end of the *bóithrín*, where they are left out, and search through the blue recycling bins for old newspapers with which I can light the fire.

In Ireland rubbish is charged by weight, meaning that every bit reused saves in a few ways. I understand the logic behind it – if people have to pay for waste, their financial self-interest should motivate them to create less of it. Reality is another thing. Instead of generating less rubbish, some people come out to places like this and dump it all over the sides of the roads and forests for free. If rubbish collection was free at the point of service, the littering of rural Ireland would cease immediately. When I go on strolls through the woods I sometimes wonder if the people who come up with these policies ever spend time in places like this.

~

After a week of lumping, sawing, chopping and stacking beech, spruce and birch, I have finally managed to get two years ahead on firewood for the first time since I moved here. It's taken me that long to reach the standard set by any smallholder worth his salt, though most smallholders these days earn that salt with chainsaws, tractors, diggers and other earth-shattering innovations. And so standing back, looking at it on a wet, darkening October evening, I'm surprised to hear myself thinking, 'I'll just get a little bit more in tomorrow morning, before breakfast.' Be careful of that mentality, Mark, I say to myself, as I put away the axe.

I notice a couple of men, with brown bags of spruce saplings in their hands, replanting the clear-fell across the field from us. It strikes me now that while machines may be unparalleled at reducing forests to numbers, it is still the intimate human hand which excels at planting trees.

~

I'm back teaching at Schumacher College in Devon, where I first met Kirsty. This time it's a week-long course with another friend, the author Shaun Chamberlin. The plan, as always, is to mess with their heads. Heads need messing with every now and again.

On the first day of the course I lay out my teaching fee for the week – £1,000 – on a table in front of the students, and tell them that they have got to decide what to do with it. It's a session on money and 'gift culture', and I've decided to make it a practical one. I offer four options, and make the case for each one in its turn. The aim is to come to a consensus, but it's immediately obvious that opinions vary wildly.

At the start of the session there is no interest in Option 1, which is to burn it all and stop the cycle of ecological violence it inflicts almost every time it is spent in an industrial economy. Some argue for Option 2, which is to give it all to a good cause of their collective choice. Most make the case for Option 3, which is to give it back to me, while the remainder want to choose Option 4, which would involve dividing it up equally among the group and letting each individual decide anonymously what they want to do with it themselves (which could simply be to go shopping with it).

After hours of heavy deliberation they decide on Option 2, and to split the money between a rewilding organisation in Wales, the Cambrian Wildwood Project – whose work is ground-breaking for the fact that it doesn't break ground – and the translation and promotion of the works of the late David Fleming, whose ideas and book, *Lean Logic*, this particular course is based upon. In the end, however, it is spared from being burnt by one vote.

In the next session, on appropriate technology, I put a sledge-hammer to what appears to be Shaun's laptop, apparently against his will. It is actually an old, broken laptop of my own, but Shaun's acting is so convincing that people think I've gone mad. The animosity towards me in the room is palpable for about five minutes, until the truth is revealed. Deep sighs of relief are heard and people start smiling at me again, though a few are still annoyed that their emotions have been played with. I ask the group how they felt in those five minutes. Many explain they felt angry since, they had believed, I hadn't had permission from Shaun to smash his computer, no matter what I thought about tech or its impact.

I asked them if they had ever experienced the same emotional response towards the corporations that make their own laptops and smartphones, considering they devastate entire habitats on their customers' behalf without any permission from the life that dwells there. No, they say, they hadn't. Not really.

On another occasion we go out with resident scientist and author Stephan Harding on what he calls a 'Deep Time Walk'. This is a 4.6-kilometre walk around the surrounding woods and coastline, with each step of the journey representing one million years of the earth's history.

Not much happens at first, as we reflect on the enormity and incomprehensible beauty of the earth's life thus far, but by the time we reach the Devon coast, halfway through the walk, Stephan is already explaining to us how life has slowly begun forming and cascading into its myriad, enchanting forms. As the ocean crashes against the cliffs below us, a number of the students are clearly having a profoundly moving experience, their own lives put into clear perspective against the spectacular expanse of existence.

The final millimetre of the 4.6-kilometre walk, Stephan tells us, contains industrial civilisation, and in that one millimetre we are in danger of wiping out much of what came before it. What the next millimetre, metre and kilometre have in store for planet earth and ourselves, none of us pretends to know.

~

I receive a letter from a prominent Irish thinker and advocate of localisation. It's typed up. Times New Roman, I believe. He starts off by saying that he had begun hand-writing it, but that it was so illegible that he could barely understand it himself. It has been a long time since I've actually had to write something, he says. He also happens to be a writer.

He's not alone. A lot of my time responding to letters is taken up trying to interpret the hand-writing of my correspondents. That a few of the letters even get here is of great credit to the postal service, as sometimes even I can't make out the address, and it's *my* address. Some of it looks like it might be written in shorthand or Arabic. Still, I prefer it to Times New Roman.

Lack of use is certainly one cause of its demise, but I've found that the first obstacle to good hand-writing is the expectation of producing forty words per minute, or the need to knock out a letter at email-speed. Once you slow down, good hand-writing becomes easier. Once you slow down, good anything becomes easier.

It strikes me, as I read his letter, that he has become entirely dependent on the machine that eroded his ability to write legibly in the first place. This pattern is not uncommon: in fact, it's the history of our relationship with technology. After all, this cheap pencil in my hand, and the even cheaper paper I'm writing on, replaced my ancestors' ability to make their own writing materials from the landscape around them.

As I write those words, my writing suddenly feels like an intellectual exercise, the act itself disembodied from the place which induced and inspired the words themselves. I'm not sure a disembodied art can help bring about a more embodied culture. But it's one step closer to home, towards which I want to keep walking.

~

Our event space is throbbing with a group of home-educated children, who are here for a free art class with Caroline Ross. But this is not your regular art class.

Caroline – who, when I first meet her, is wearing moccasins made from buckskin she skinned and tanned herself – was inspired by Paul Kingsnorth's first novel, *The Wake*, a post-apocalyptic story set in 1066 (until I had read that book I had never thought of history as a long series of apocalypses before). After reading *The Wake* herself, Caroline wondered what art would have looked like in the days of Edward the Confessor. Back then they had none of the commercial pens, brushes, acrylics and oil paints that artists use today, so what did they use?

She begins the lesson by telling the children about pigments that have survived for five hundred years, along with stories of ochres on cave walls that are fifty thousand years old. She shows us – I'm as keen as any of the children – how to make a goose-feathered quill pen. Being left-handed, it takes a bit of practice not to smudge the ink across the paper, but it works perfectly well otherwise. I wonder if it is because of these original pens that so few of us were, and still are, lefties.

We make pencils from small sticks, sharpened at one end, which are then dipped in ink. But this is not just any regular ink, either. She has pigments of green earths from the Lake District, yellow sinopia from Oxfordshire, red ochre from the Forest of Dean. To make coloured paints, she cracks open an egg, before whipping the whites and waiting for the liquid to settle underneath. This she mixes in a mussel shell with some of the ground earths. They make a brilliant paint, the kind used in the old, illuminated manuscripts.

She shows us mushroom paper, and ink made from ink-cap mushrooms; there's gesso, a mix of chalk, animal glue and white pigment, with which artists in the Renaissance prepared canvases; rich black colours made from Welsh oak galls and rusty nails; she has even created her own pencil cases from birch bark stitched with deer sinew, while the pouches for her other materials have been crafted from fish skins.

Caroline, it seems, is not merely drawing and painting landscapes. The art is itself a part of the landscape.

By the time she leaves, the children are painting like it's 1099, and most of the parents feel like a whole new world has opened up to them. There's only one problem, from my perspective: despite all of this greater understanding, I still can't draw.

~

My neighbour Kathleen asks me, Kirsty and Brian (who lives in the farmhouse) for a hand bringing in the wood. She has two huge piles, the result of an order from the local council which told her that some of the trees in her small woods were a safety risk to drivers. It only takes us an hour to stack them under cover. As always, she tries to give us money when we're done. A tussle ensues, and eventually we accept a few vegetables from her garden, along with lunch with her and her husband. Packie once told me that, when he was growing up, everyone 'put down the vegetables'. Of that generation, Kathleen is now the last person around here with a vegetable patch.

Over lunch she tells us that she has just been to London to visit
her sister. She goes there once a year. Her husband Jack tells us he
has been to England once, and Dublin and Galway a few times
too. Kathleen's sister left when she was young because, as Jack
said, there were no jobs here, and by that stage jobs had become
part of the Irish psyche. Until then, 'work' and 'a job' were two
very different propositions.

On her way back from Dublin this time, she lost her passport.
She tells us that she has tried to get a new one since, but that
when she phones the passport department to sort it out she can't
manage to speak to anyone; it just asks her to press #3 for this or
#4 for that, before finally instructing her to visit 'w w something-
or-other', but she 'hasn't got a clue about any of that craic'.
While she's in the mood she says that she's seen shops in London
where you check out your own shopping. Looking disturbed, she
says, 'But sure you can't chat to a machine, can you?' Actually,
you can now, I think to myself, but I say nothing as I'm not
convinced that a vision of AI cashiers will make her feel any
better.

~

November is not such a busy time in the garden, but there are still
some jobs that need doing.

First up today is a box of garlic cloves that need planting, which
we saved from our previous crop. Garlic gives better interest rates
than any bank I've ever known. Bank a clove, and you reap a
whole bulb. A six hundred per cent return, sometimes. I find it's a
sound investment, and one less subject to the vagaries of events in
London or Tokyo. The dividend will come in around July, some
of which will be reinvested.

The beds need preparation, so that the impending heavy frosts of January and February can start to break things down. Conventional growing wisdom advises people to dig and turn the soil over, but we take a no-dig approach to gardening wherever possible. We haul wheelbarrow after wheelbarrow of manure and compost onto the raised beds all day. This way the delicate web of life in the soil is left undisturbed, leaving the earthworms to do the hard work for us between now and spring equinox.

There's some old fencing that needs to come down. I'm always happy to take down a fence. I know well their use in terms of enclosing animals, and I erect them where absolutely necessary myself, but I've yet to look at a natural landscape and feel it would be improved by a fence.

~

I can't remember the last day I've had completely off. In this way of life, a 'duvet day' could come close to killing you. There's none of the convenience of taps to turn, buttons to press, automated central heating timers to set, cafés to pop into or switches to allow you to put your feet up for the whole day. There is always something. Always.

The flip side to this is that, most days, I feel absolutely alive.

~

I've had a shaved head – a number one – for most of my adult life. I used to shave it myself, once every week or two, with an electric razor, and I remember having always felt sharper, neater, more efficient – more 'on it' – for a few days after each shave.

A few months before I unplugged, I found a 1960s set of manual hair clippers in an antiques shop, the kind sold for use on horses today. The owner of the store, which is a great source of obsolete, inexpensive, quality hand tools, told me that they work just as well as the modern electrical version, except that they are a little slower and that it's more difficult to do the back of the head by yourself. I've yet to find out, as they've been sitting in a drawer ever since.

Twelve months later I now have a full head of hair and a beard. I've found it a peculiar experience, touching on something more profound within me that I hadn't anticipated. I tend not to look in mirrors – horrible, self-analysing, vanity-inducing little things – but on the rare occasion I catch myself in something reflective it can feel strange to see the person looking back, like I don't fully recognise him yet. Neither do old friends who haven't seen me for a while. One suggests that my new haircut fits the lifestyle I'm living better – the image of the savage, wild, uncivilised man we have in our minds – and while there might be some natural reasons for that, I certainly don't want to start branding myself.

Why have I not used those manual clippers? I'm not sure. I've thought about it a few times. Maybe a part of me is wanting to let go of an old sense of self – that sharp, neat, efficient guy who was more at home in the city. Or maybe I've stopped shaving for the same reason I no longer want to scythe every square inch of the land. Or, maybe, I'm finally learning that none of this stuff is important, anyway.

~

Journalists and visitors often ask me some variant of the question: what do you think will be the most important skill of the future?

I don't know, I tell them. If the futurist Ray Kurzweil and his buddies at Google are right, it may be robotics engineering. If, on the other hand, the climate scientists and ecologists (or those rare economists who understand ecology) are right, it could be any of the things that the great mass of people can no longer remember how to do.

The techno-utopians are putting all their eggs in the AI basket, but no matter what the future holds I know the kind of life I want to lead. Give me natural intelligence over artificial intelligence any day.

~

Before moving back to Ireland, I lived on an organic farm among the rolling hills between Bristol and Bath in England. Johnny Depp was said to be living around there too, but Johnny Depp is said to be living everywhere. Good for house prices.

On the surface, the cultures of Ireland and England are similar, and are becoming more homogenised, as is the way with globalisation. But underneath the superficial layers of their identikit cities and large towns there are still distinctions, and these, I find, become more pronounced as you move out towards the small villages and parishes of the countryside.

Having spent considerable time in both countries, I've noticed many differences between rural England and rural Ireland. Some of these differences are blatant, the most obvious being house and land prices, and their effect on the rurality of both countries (Ireland's countryside being largely populated with small-scale farmers, England's with agribusinesses and commuters). With Ireland's city-educated twentysomethings flocking to the urban centres, each following the other in pursuit of careers, excitement and bigger money, the cost of putting down roots in rural Ireland is a fraction of that in the English countryside. At a time when the UN is declaring that the internet is a basic human right, the most basic right of all – to build a simple shelter where you can feed yourself and your family – seems to be drifting further out of reach than ever there. And pretty much everywhere.

But I also noticed more subtle differences. The English work less human-scale farms, create better hedges, don't chop down all their trees, and generally don't sit at the bar in pubs. But the differences run deeper too, into the very make-up of our psyche. One of the first questions people in rural Ireland usually ask, when meeting you for the first time, is 'Where are you from?' In rural England, the people you meet are more likely to ask 'What do you do for a living?' I have always found that an awkward, unattractive question, even during those days when I was living what might be considered a more conventionally successful life.

At a talk I was giving a few days ago, about my life without technology, the host introduced me as 'a writer'. I didn't correct him at the time, as I didn't want to make a fuss. Yet I felt uncomfortable about it. I had never set out to be a writer, and still don't consider myself one, at least no more so than I am a woodsman, grower, fisherman, forager or any of the hundred other things which get me from one end of the year to the next. At most, writing is a by-product of my actual livelihood, and something I do for reasons other than putting food in our stomachs. But I suppose 'writer' sounds better, more creditable, more intellectual, especially as one is about to give a talk.

In a world where those who do the least amount of work (the shareholders of multinational booksellers) make more money per book than those who do the majority of it (the writers), I've certainly found the axe to be mightier than the pen.

~

For all of my working life, diaries have organised the minutes and hours of my days. They have ensured I hit deadlines, sent emails or met friends on time, scheduled meetings, paid bills, got on trains, done to-dos and a thousand other small, indispensable things.

I haven't looked at my diary now for two months. My work is in front of my eyes, though some days I would rather it weren't.

I still can't decide whether I'm losing touch with reality, or finally finding it.

~

After coming back from a sabbatical to New Zealand in 2006, I decided never to fly again. The decision to stop flying wasn't easy. I had enjoyed travelling, and looking down on human civilisation and expansive landscapes from another perspective. I won't go into the reasons why I stopped flying; by now we all know them only too well, and yet we fly more than ever. Business executives attend important meetings in Dubai, tourists spend weekends in Amsterdam and Berlin, hippies do spiritual retreats in India, while environmentalists hold international conferences in close proximity to international airports. It's just what my generation does. We expect it, and it is expected of us.

For six months I've known that I have to go to Norfolk, in the east of England. I have three reasons: a birthday party, a fortieth anniversary, and to spend time with Kirsty's family, who have lived there all their lives. All of these things feel as important as my reasons for rejecting technology; but important things can often come into conflict. My options for getting to Norfolk are limited. Sailing isn't yet one, and so from this position onwards the inevitable compromise is merely a matter of degrees.

We board the ferry at 2 a.m., hoping to get a few hours' sleep on one of the sofas in the lounge. We'll be lucky, as there's a group of students on the piss and the alpha-male is loud. We're eleven hours into a twenty-six hour journey, one that is forty-five minutes by air. People fly from Ireland to Australia quicker. I understand why people fly. My friends think travelling over land and sea for twenty-six hours is extreme. As I look around this epic, aquatic monster and its cinema, shops, restaurants, amusements, bars and floors of accommodation, I would have to agree.

The next day, we're in the toilets of a service station, where the driver we've just hitched a lift with is taking a break. The complex looks identical to every other service station I've ever been in, with the same six corporations providing the hamburgers, chocolate, coffee, cigarettes, accommodation and news. There's no indication, anywhere, of what part of the country we are now in. A cover version of Joni Mitchell's 'Big Yellow Taxi' is, rather absurdly, being piped through the airwaves as my piss disappears downwards out of a spotlessly clean urinal. It isn't paradise, but it does have a parking lot. I had thought about going under the mature, lonely-looking oak tree next to the car park, but there were people everywhere, people who may not have appreciated me pissing there. When we get back in the car, the driver thinks we're near Bishop's Stortford, but he doesn't know for sure either. M11, he says.

As soon as we reach her old home in rural Norfolk, Kirsty's father David is pouring us both a large glass of plum wine which a friend of his had made. What a mission, he says. David has become a close friend of mine, and the length of the journey has made seeing him even more special. By the time we finish the second glass, we're ready for bed.

~

Setting the fire, part III.

A story on the front page of a regional newspaper I'm about to burn says that the government has just released a report claiming that 'the blue economy' – what poets and romantics might still sentimentally refer to as 'the ocean' – could be worth €4.7 billion to the Irish economy; that is, if we can build up industries capable of exploiting it to its full potential. Energy production, tourism and all sorts of other initiatives are mooted as ways of creating new jobs for the west of Ireland, boosting economic growth, and opening up new opportunities for graduates. The blue economy is the next big thing. Every square metre of under-utilised ocean is basically lost jobs, lost money, lost opportunity. Who could argue with that?

I've heard the same argument made about the Arctic, and every other wild place on the planet.

~

'Twenty years a-growing, twenty years in bloom, twenty years a-stooping and twenty years declining.' While Muiris Ó Súilleabháin himself was just beginning to bloom, the island around him had begun to stoop. Having endured the Great Famine, many personal tragedies, armed bailiffs and daily battles with one of the most dangerous stretches of water in Europe, the people of the Great Blasket finally met their match in the form of a ruling triumvirate who had, long since, been gathering steam elsewhere: globalisation, mass urbanisation and their father, industrialism.

Though their livelihoods comprised a thousand little things, only fishing brought them any real financial reward. And for their first 130 years there they earned enough to get themselves from one year to the next, with the sale of fish paying for salt, dowries, boots, coffins and breeches along the way.

Such was the abundance of mackerel, pollock, cod and lobster – coupled with their small boats' inability to overfish – that in 1921 there were four hundred naomhóga *fishing off the West Kerry coast, all providing decent livelihoods for fishermen whose material needs were quite simple. Times were more or less good, and the population of the island was at its peak at around 175 people (which, interestingly, is around the number of people that anthropologist Robin Dunbar suggests we can maintain a meaningful relationship with). But just beyond the horizon, industry was on its way, and by the summer of 1921 change was floating in on the breeze that came off the Sound. In* Island Cross-Talk, *Ó Criomhthain begins to see the dangers ahead:*

> *When I turned to gaze north, there I could see boats fishing away like any other day of the week, but they were not from here, they were from the land of France. They are causing great harm and they will cause more, and not in one way only. Besides carrying off the fish, they are weakening the Faith too, for the poor Island fisherman is watching them catch the share of the fish that should be his, on a Sunday, which for him is a day of rest.*

A few years later, Ó Súilleabháin would prophesy the future of the island's people with remarkable precision. Speaking with a girl he had just fallen for, Muiris proclaimed:

> *Don't you see it yourself? The chief livelihood – that's the fishing – is gone under foot, and when the fishing is gone under foot the Blasket is gone under foot, for all the boys and girls who have any vigour in them will go over the sea: and take the tip off my ear, Mauraid, if that day is far hence.*

The year after these words were published in 1933, the local naomhóga *fishing off the coast of Kerry numbered no more than eighty, and with the falling prices that are heralded as one of the many benefits of industry, even the Islanders struggled to keep going.*

With the strength and faith of the people weakened, the industrial machine – in its need for efficiency, labour and markets – began its second wave of attack on the Great Blasket: mass urbanisation, and the promise of a better life elsewhere. Large ships began taking people on the long voyage to America. By 1947 Gearóid Cheaist Ó Catháin would be the youngest person on the island by thirty years and thus earned international fame as 'the loneliest boy in the world'. He wrote in his memoirs that once the exodus of young people had started, 'those who remained had no choice but to move'. Most headed for the next parish west – Springfield, Massachusetts, USA.

The night before their sons and daughters would leave, the Islanders would have an 'American Wake', as those who made the long journey would likely never return, nor be seen alive by their own kinsfolk again.

The exodus was understandable. Life on the island was tough, and America was presenting itself as a land of plenty for those willing to work hard enough. Unlike the back-to-the-landers of the 1960s, who had experienced urban life and were consciously rejecting it, the young Islanders at this point could only see the exciting possibilities, and knew little of the sacrifices that would eventually come with the new life. In his foreword to the poet Micheál Ó Guithín's book, A Pity Youth Does Not Last, Professor George Thomson writes that, 'America offered those exiles an escape from the poverty of their home life, but only on the condition that they surrendered their cultural values. Some paid this price without regret, others with a life-long sense of loss; for a few it was too heavy, and they came home.'

Ó Guithín was one of those for whom the price was too heavy; but by the time he returned, the island was on the brink of evacuation. He would later mourn the death of his home, and wrote in his poem, 'A Rock, Great its Fame':

> Each man set off for himself
> The panic was senseless
> From that day out truly
> Friend did not stand by friend

Looking at my home here in Knockmoyle, I have a sense of how Ó Guithín may have felt. I can see the same forces at play under my own nose, and the same slow death unfolding. The local pubs know it too. As Ó Súilleabháin's grandfather would say, it's twenty years a-stooping. The population here is getting old, and farmers' children have no interest in continuing the farms that their families built up over generations. Though the tractors first brought ease and speed, they soon replaced the need for labour, and so the labour – otherwise known as young men and women – took off to cities to work in factories, offices and the service industries, and made new lives for themselves there.

I'm reminded of Aldo Leopold's words: 'In the long-run too much comfort only seems to spell danger.'

~

Out for an evening's walk, Kirsty and I meet Aisling from up the road, who is out doing the same. We stop to talk. She had seen us out on the horse and gig a few mornings ago, as we were collecting a few sacks of sawdust for our composting toilets from a nearby, seldom-used sawmill. She says the sight of it took her back fifty years.

She tells us that, when she was young, she would take the horse and cart the 6 kilometres to the shop, every Tuesday, to pick up the week's flour, along with any other groceries her family needed, and could afford, if any. The trip took all day, but only because they called into all their friends and neighbours along the way. Hard enough times, she said, but happy times.

With that she's off, power-walking down the *bóithrín* in her high-visibility bib. We stroll off in the opposite direction, where we need to water the horse.

~

The postmaster tells me that it has been announced that four hundred of the remaining eleven hundred post offices in Ireland are to be closed, and that all of the closures will be in the less-profitable rural areas. Right now he doesn't know if his own office will be one of those getting the axe. Nobody knows yet. No point worrying about it, he says stoically, as there's not much to be done about any of it. It's a decision that will be made at headquarters, in Dublin.

If this office does close, it will mean an extra 24-kilometre cycle for me to and from the nearest town, but my troubles will be hardly worth mentioning in comparison to his, or to the locals for whom the post office is more than just a place to pick up their pensions and send parcels to far-flung family.

He tells me the week's weather forecast too. It's not good. There's a hurricane on the way, he says, so get yourself prepared. Next to the cash register in the adjacent shop there's now a large box of loose candles, already half gone, in the place where the chocolate bars have always been before.

~

After much hullabaloo, Hurricane Ophelia finally hit the west coast of Ireland today. Neighbours told me that the media were predicting that it would cause €700 million worth of damage. That's actually good for the economy, I tell them; or somebody's economy, at least. Universities, schools and shops have closed down for the day, and even the postman isn't delivering this morning. Such is the collective fear of the nation that the electricity has blacked-out before the event. Wind speeds of 140 kilometres per hour, they say.

Ophelia reaches Knockmoyle by mid-afternoon. She's in a right old mood. Who can blame her? I grab my coat and go for a walk. Good to feel the wind in your hair. I need the elements. They help to keep me in my place, save me from any delusions of grandeur and remind me that I need to appease the gods of water, earth, wind and fire. A few trees have already come down, her vengeance brutally indiscriminate. The roads are empty. Nice. The chickens are hiding in the coop, which we've secured in the lean-to. Ophelia roars, and I try to listen.

Those in the farmhouse come over to the cabin for dinner. As we don't have plumbing, central heating or a boiler, we're able to put down a fire without the risk of anything exploding. The electricity may be off for hours or days. It's more likely to be the latter, as resources will be thrown at the cities first.

Jorne, who has spent over a decade captaining sailing ships from Europe to the Caribbean, explains to me that as the temperature of the oceans increases, hurricanes will become more prevalent. Best get used to it, he says.

~

In his essay 'Axe-in-Hand', Aldo Leopold wrote:

> The Lord giveth, and the Lord taketh away, but He is no longer the only one to do so. When some remote ancestor of ours invented the shovel, he became a giver: he could plant a tree. And when the axe was invented, he became a taker: he could chop it down.

After a season of giving (spring) and a season of taking (summer) I'm now due a season of giving again. November to me means tree-planting time, and I take whatever opportunities I can get, my ambitions limited only by the amount of land I am entitled to do it on.

Kirsty's family own land which is farmed industrially – sugar beet and crops intended as animal feed – and after a few years of conversation we agree to plant a small woodland on a four-acre section of their land. It's a small gesture towards ecological restoration and yet, at the same time, a bold and brave move, running in the face of conventional wisdom regarding how farmers use so-called agricultural land. Most farmers would grow any crop other than trees on their land, even if it is one – such as sugar – which is proven to be detrimental to human health. Anything but trees. One of her family tells us that the woods we're planning will be financially unviable. I tell him that I hope so, and that his surfboard and music system are financially unviable, too.

After much advice from a good man in the Forestry Commission and various ecologists, we decide to mirror the adjacent, existing woodland and the species which comprise it, as considering its age – some of the trees are over four hundred years old – these are the varieties best suited to this particular place. We will plant in clusters, according to the needs of each species, with forty per cent of the patch kept as a meadow for wildlife and wildflowers.

We lay the trees out. Pedunculate oak, silver birch, downy birch, field maple, hazel, holly, hawthorn and dog rose (Kirsty introduced the dog rose, which wasn't growing in the existing woodland, simply because she finds it beautiful, which was a good enough reason for me). Sixteen hundred trees in total. They each come with their own guard to protect the young trees from rabbits, which are abundant here due to the absence of predators, something that's a trademark of vast agricultural landscapes.

I'm torn over the guards. On one hand they're made of hard plastic. On the other hand they'll protect the trees from rabbits which, because of supermarket chicken, beef and soya, have become too inconvenient to kill, skin, butcher and pot up anymore. My own ideal is to simply leave the land alone and let it rewild itself, as it has been doing this very well since the beginning of time. Saplings grow up fast through the pioneering brambles, which naturally protect the trees from deer and rabbits and provide food for wildlife. The trees, once they begin maturing, eventually shade out much of the brambles, and in time a native woodland establishes. It is only such self-willed woods as these that have any hope of becoming ancient.

The idea of allowing the land to rewild itself is still controversial in farming communities, however, and so our options were restricted to it being industrially farmed with insecticides, pesticides and fertilisers, or for us to plant a so-called traditional English woodland instead.

Planting a woodland can be as simple or as complex as you like, depending on whether you opt for a gridded plantation or something resembling a native, natural woodland. If you veer towards the latter, as we have, the main work is in understanding the land, and listening to what it might want. Putting the trees in the ground is the easy part.

The technique I use for bare-rooted trees is one called slit-planting. This involves little more than shoving a spade in the ground, pushing it forwards to make a slit and dropping a young tree into the opening, before firming matters up. A child could do it. A child should do it. I can plant about 250 trees a day when planting this kind of woodland (which is much slower work than a gridded plantation), but a good forester can do upwards of 500.

Re-reading Leopold, I reflect on the spade as a tool. As well as planting trees, it can also be used to disturb soil, harming the life within it. Like all technologies, even the laptop, it can be used to give or take life at a greater pace than we would be capable of ourselves. But Paul Kingsnorth captured the differences in these technologies best when, in *Confessions of a Recovering Environmentalist*, he wrote:

> Both may give you sore arms, but there is a difference between a keyboard and a spade. A spade can still be made fairly simply. It doesn't need constant energy to keep going. It can last a long time, if you treat it well, rather like your body. A keyboard and a spade are both products of an industrial economy, but not to the same extent, and they do not have the same purpose. One can exist independently, the other cannot. This might be a matter of degrees, but the degrees matter – and so does the intent.
>
> There's another point too, though, and perhaps it is a more important one: nobody ever got addicted to a spade.

Because we're not planting in straight rows, the woodland takes three of us more than two days to plant out and protect. It's satisfying work, even in a cold northerly breeze that gets in at your bones if you stop for a break. Over the course of the whole weekend, I notice three earthworms. Three. Looking across the field, all I can see yet are green upright plastic tubes, but I hold tight to the hope that in twenty years' time I might walk through this field and witness the return of life – trees, wildlife, insects, wildflowers and earthworms – all carrying out the age-old task of feeding each other and creating life through death.

In 'Axe-in-Hand', Leopold tries, and fails, to explain why he prefers pine trees to all others. In the end, the best he can do is to say, 'The only conclusion I have ever reached is that I love all trees, but I am in love with pines.' Well, I love all trees, but I am in love with oaks.

~

Some of the neighbours have caught wind of the fact that we make our own cider, and so during the first week of November sacks of eaters and cookers continually land at our door. While none of the natives use them themselves any more, the older generation still hate to see them go to waste. Their sons and daughters, who would have traditionally provided the muscle to press apples, are all off in Dublin, Toronto, London and Sydney, earning good wages so that they can buy things like bottles of cider.

I go through each apple, separate the good from the bad, and from the latter I cut out any rotten bits, put them in a crate, and pass them to Kirsty. She puts these through the old, rusty Wexford self-feeder which is at least one hundred years old and which the previous owner had used to grind up swedes for his pigs. With these all chopped up in a bucket, she passes the baton to Jorne, who hand-cranks them into pulp so that Elise can press every last drop of juice out of it. From there it goes straight into a barrel – no sugar, no yeast, just apples. It's almost like a factory system. Almost.

By lunch, we've all had enough of apples for one day. But it has been a fun, rewarding morning in which our lungs were full of clean, crisp air and work song. There will be 46 litres of cider in the barrel – a shrewd businessman would value this at €230, retail price. Our first job, however, is to drop around a few bottles of apple juice to the local 'pioneers' – those who, aged twelve, took a pledge at their confirmation to never drink alcohol. The rest – the majority – will have to wait six months, like ourselves.

~

Kirsty likes her eggs soft-boiled. Without a watch, they're difficult to get just right. So I take six eggs – four for myself, two for Kirsty – out of the pot, and hope that I haven't left them in too long. I peel the shells into a mortar and grind them with the pestle. I've read that ground-up eggshells, being high in calcium, can aid the regeneration of your teeth. Considering that I grew up on chocolate, sweets and fizzy drinks, I'm keen to give it a go.

They're more edible than I had imagined, but that's not saying a whole lot. They taste of, well, eggs – which, surprisingly, I find surprising – but I can't see them appearing on restaurant menus any time soon. Still, I prefer them to buying tubs of calcium supplements from the health food store I used to run.

~

You can really feel the change of seasons today. I woke up not so much feeling ill, but not feeling full of vitality, either. Life has been busy of late. I find that it's the communication of this way of life – the writing, talks, interviews and curious visitors – that tires me; the life itself only seems to keep me in good health, physically and mentally.

Kirsty makes me a pot of tea from herbs – red clover, silverweed, raspberry leaf, calendula and chamomile – which she picked and dried earlier in the year. Such teas are not intended to treat symptoms directly, in the way that we use industrial medicines; instead they aim to aid the body in the task of healing itself, something it always wants to do. While the tea is brewing, I chop and eat five cloves of raw garlic. With that I decide to put away the pencil, light the fire, grab a book, kick back and take the afternoon off.

If you don't make time for health, you have to make time for illness.

~

Like most of our dinners, today's is common fare: potatoes, swede and garlic from the garden, roasted in an oven heated by spruce and beech, alongside a good-sized jack pike from the lake and a few sprigs of rosemary and thyme from the herb garden. To go with it I pick a bowl of salad: kale, pak choi, rainbow chard, purple sprouting broccoli leaves, rocket, mustard lettuce and parsley.

If we had kept the polytunnel that, against conventional wisdom, we took down to build the cabin, I've no doubt that we would have a few more varieties of vegetables at this time of year. I was under no illusion otherwise. But I'm satisfied with the choice we made to eat an Irish diet – what this climate can provide without recourse to violent products like polytunnels – and all that it entails. Sometimes I miss peanut butter, bananas, halva, olives, sun-dried tomatoes, cashew nut butter, hummus and other delights that can only be grown in other climes, but mostly I don't. And there's a sense of real security that comes from knowing that, no matter what crises or catastrophes unfold in the wider world, you know how to put food on the table for yourself, your neighbours and those you particularly love.

And then I remember that damned monofilament line which helped me catch the pike. I have books full of primitive fishing techniques – Ray Mears and John 'Lofty' Wiseman being the most useful – but there are two problems with them. One is that they are all illegal in an industrial world that at the same time not only permits, but actively encourages, bottom-trawling. Two is that they were effective at a time when our rivers were heaving with fish. Our rivers and lakes are now as barren as the soil that, along with insecticides and herbicides, washes into them with every heavy rainfall.

One day, our waterways will be clean and full of life again, I tell myself in hope, and it's a day I'd like to see before I die. But I fear that if we don't learn from our past, and our mistakes, it may not happen until after many of us are gone.

~

Speaking of which, I've just walked past the chapel in one of our nearest towns. It's a glorious day, and four of its devotees are out tending to its garden. When I say 'tending' what I mean is that they are spraying herbicide on its manicured lawn. The man who is doing the actual spraying is walking along the bank of a river, and I notice that some of the herbicide is being sprayed into the river directly (more will inevitably find its way in on the next heavy rain). As he turns around to spray the stretch of grass that runs parallel to the church footpath, two of the other men – who, up to now, had been spectators – suddenly spring into action and walk alongside him, holding up a large PVC board to prevent the spray leaving an ugly film on the tarmac which leads the congregation into their chapel. I ask them, in a friendly manner, if they could at least not spray next to the river. One of them tells me to 'leave us in peace'. I assume to love and serve the Lord. The Lord must prefer tarmac to rivers.

~

On 23 November 1953, the last of the Islanders were evacuated from the Great Blasket.

Thirty years later the translator Tim Enright would say:

One cannot mourn the ending of a way of life that, especially in winter, was very bleak indeed; one cannot but mourn the ending of a culture that was rooted in the far distant past.

~

There are three traditional methods of dressing a deerskin: brains, eggs and soap. Modern tanneries usually use chromic acid, which is cheap but toxic and cuts swathes through life, devastating the waterways surrounding tanneries in countries with lax environmental laws. According to Matt Richards, in his comprehensive practical guide *Deerskins into Buckskins*, these tanneries then market the chrome-tanned deerskins as buckskin, despite it having 'very different properties than the traditional material'.

Strangely there's just enough brains in an animal's skull to tan their hide, so I do it that way. I cut away the skin between the deer's eyes and his antlers, before sawing a v-shaped open-ing into the skull until the brains are exposed. I put my fingers into his skull and scrape out the brains and put them into a bowl of hot water, before mutilating and liquefying the brains into a soup.

Into this I soak the dried deerskin – which has already been fleshed, soaked in a wood ash solution for three days, grained, rinsed and membraned – and leave it for half the morning, before I wring it, dress it and wring it again. This morning I don't have time to soften it – the process which made buckskin such a valued material for millennia – so I stretch it out a bit and hang it up to dry.

~

I'm excited. It has been a long time since I've had a hot, relaxing bath after a hard day.

I filled the bath with water earlier in the day, and covered it with a lid I made out of clear, corrugated Perspex, which we took off the old pig shed. This allows the afternoon sun to take the chill out of the water, meaning I need a bit less wood to heat it up. Before I set the fire I swap this lid for another wooden, insulated one, so that when the water does start warming up, the heat won't drift off into the dark night sky. When you haul, saw, chop and stack your own wood by hand, you take care to use it as wisely as possible. Washing ourselves has been the most challenging aspect of the last eleven months, but the hot tub before us looks all the more appealing for it.

Kirsty has a quick rinse on the wooden decking to wash off the worst of the day's dirt, so that the bath water stays largely clean for those who come after, and she gets in first. A woman's prerogative. She lights a couple of candles as I pour two goblets of blackcurrant wine. I get in slowly, my body readjusting from the chilly November air to the sudorific water. That's more like it. Lying back, the heat of the cast iron radiates through my body, teasing its way into every vertebra, tendon and ache, while a gentle drizzle refreshes my face. On reflection, I should have prioritised building this hot tub at the start of the year, but then again, I had a lot of things to prioritise when I first started down this path. But none of it matters now.

~

Late autumn is when I start thinking about manuring the vegetable garden. If the soil isn't full of nutrition, it's impossible for the vegetables to be. If the vegetables aren't full of nutrition, it's impossible for our bodies to be.

We collect around ten wheelbarrow loads of horse manure from a stables up the road, and heap it in a mound. A couple of days later, an experienced organic market gardener we know notices it, and warns us to be careful of a wormer which equestrian folk routinely treat their horses with, as it will still be in the manure and get taken up by the vegetables in the same way as its minerals and vitamins are. Ultimately, he says, it will end up in your body. We check with the stables, and he's right, it is in there, and lots of it, too. The whole affair reminds me why I stay clear of what we consider to be 'normal' food. Kirsty has recently treated the horses she looks after for Packie with carrots and garlic, which is the old traditional treatment. She is off on an adventure soon, and wanted to make sure that they were in good condition before she left.

Despite it being more labour-intensive, we decide that from now on we'll poo-pick the field in which the horses are kept. That involves pushing a wheelbarrow around a muddy field in November and hand-picking their shit off the grass. The benefits, however, are twofold; we get to eat food that doesn't have wormer in it, and the horses get to eat food that doesn't have their own shit in it, lessening the likelihood of them getting worms in the first place. Wild horses never needed to be wormed before we started 'looking after' them, but wild horses weren't stuck for months on end in a two-acre field with nothing else to eat but the grass under their own shit.

It would be one of those win-win-win situations if it didn't involve me pushing a wheelbarrow around a field picking up shit for half a morning every day for the next month. So for now it's just a win-win.

~

Down at the lake. I'm fishing for pike, but I catch a salmon. The first of my life. She puts up a terrible fight, but I finally get her out, take out the hook, and have her in my hand. What to do now?

The law says it's November, out-of-season time for salmon, so throw her back in. My belly says it doesn't care what month it is, I'm hungry, so kill her. My eyes look into hers, I see her wild spirit and think about what my kind has already done to hers, and my head says throw her back in. The animal in me says stop being so civilised, it's flesh – do you think bears care that it's salmon spawning season when they catch them going upstream? – so eat her. My head says that bears don't dam (and thus damn) rivers and pollute lakes, so throw her back in. My belly reminds me that it is still there, and hungry, and that ultimately my flesh depends on hers. My hand feels the pulsating, magnificent life in her body, and wants it to continue going forth and multiplying. Considering that it's a once-in-a-lifetime event for most fishermen, these days, most people would suggest at least taking a photo of her. I can't, and wouldn't even if I could. No need to add insult to injury.

I look into her eye again, and I throw her back in. Not because of the law, but because . . . well, I'm not sure actually. For some reason unknown to my stomach, it just didn't feel right.

I cast in again, and unlike every other fisherman in the country, I hope for pike this time instead. Best to keep these things simple.

~

On my way to a babysitting evening at a friend's house, I take a quick detour to the secluded pool on the Cappagh river. Excitedly I walk through fields of grass to this particular spot in the universe, a place so worthless, so unviable for farmers, that it is allowed to burst at the seams with life.

Did I say *is*? *Was*. As I approach the pool from the north, I soon realise that this place I love is now as good as dead. The mature trees, shrubs and plants which once cloaked its banks, and provided cover for a heron, kingfisher, ducks, pike, trout, salmon and an entire micro-world of interdependent species, have been ripped out with a digger. In its place is a new farm track running adjacent to the river, complete with a barbed wire fence for its full length.

I've seen more clear-fells and strip mines and factory farms than I care to remember, but few have hit me as hard as the change to this river. I've never seen any sight so savage; actually, I've never seen any sight so civilised. It feels worse for the fact that this little sanctuary for wildlife held a special place in my own heart, and that it was all so utterly unnecessary.

I've recently been reading the ecologist Pádraic Fogarty's book *Whittled Away*, which explores how, despite Ireland's eagerness to confuse its association with all things coloured green with a global brand image of sustainability, nature in Ireland is vanishing at an alarming rate. As I stand on the bare banks of the river, a picture tells a thousand words.

Did I say the river was as good as dead? No, the river will come alive again. All we need to do to help it is nothing – which seems to be the hardest thing of all to do.

~

It's commonly thought that, living as I do, December must be tough. In some ways, there is an element of truth in this, but only if you don't enjoy the elements. In other ways, December can be the easiest month of the year. The hard work has been done. The wood has been gathered, sawn, chopped and stacked; it now only needs to be sat by as it burns. The fruit has been preserved, the herbs dried, the venison smoked, the blackberries fermented, the skins tanned, the winter vegetables planted; they now only need to be enjoyed. There are still things to be done – rain, hail or shine – but the long dark evenings do their best to attune you to their rhythm, resist as you may.

This has been my first autumn without electricity, meaning no screens, no push-button connectivity with loved ones, no bright lights to encourage 24/7 ambition, nothing to distract me from myself. Some evenings my mind will drift to old friends, people I once knew well but whom, after they and I had scattered ourselves around the world in this most transient of cultures, I've not set eyes on for years. In recent years I stayed in touch with them via email, phone, online video calling and social media. I miss them, and a few in particular. Many of them I may never see again, never even hear their voices. They could get married, divorced, have kids, get cancer, win the lottery or die, and I would probably never hear anything about it.

There are moments, like now, when I grieve for that reality. And then there are others, when I smile at the knowledge that they're out there in the world and that one day, if the conditions are favourable, one of us may surprise the other.

~

I receive a letter from my editor at the *Guardian*. Among other things, he tells me that my last article was their most shared on social media over the course of the week it was published. He tells me like it is good news – why bother writing after all, if you don't want as many people as possible to read it? – and at all other points in my life it would have been, but . . . but, somehow it no longer quite feels that way anymore. Success now seems to mean other people staring at a screen a little longer, 'liking' you, sharing your work on the websites of shadowy Silicon Valley billionaires, and I can't say I'm comfortable with the thought.

~

There's a sign up in the post office saying it is closing down. There was once a post office in Knockmoyle, just a short walk from where I live. Now the people of this rural community are going to have to travel to Loughrea to send a small parcel or pick up their pension.

I ask Packie, who is one himself, if pensioners can get their money paid straight into their bank account instead. Of course, he says. But he tells me that many of them don't understand any of 'that bank stuff'. And that you can't have any craic with one of those card machines.

~

One of the neighbours' tractors won't start. This is hardly surprising, as it has no battery, nor has it had one since I've lived here. It's not the only problem the tractor has. Instead of lights, there are four fluorescent jackets hung on the front and back corners. All of the windows are devoid of glass.

In the spring to autumn months, he parks it on a hill and starts it by free-wheeling down the track. It's a technique he has mastered. On freezing December mornings, like this one, it doesn't work so well. It's frosty today, so when he knocks on my door I'm already expecting him. He's got a screwdriver in his hand, ready to go. As he uses it to spark some part of the engine, I go against everything I was ever told about fuel and engines and light a piece of old newspaper and stick it into a vent. The flame warms the engine up, and within a few moments he's in the cab of the tractor and giving me a wave.

~

I find it strange that, in a world where free speech is so coveted, what I long for most is the freedom to not have to speak. Aldo Leopold summed up my morning's sentiments when he said, 'Of what avail are forty freedoms without a blank spot on the map.'

I take out my Ordnance Survey map, number fifty-two, in the hope of such a spot. There's nothing resembling wilderness, but I'm pleased to discover a handful of grid squares – each 1 kilometre squared – which are devoid of any of the tiny black squares that depict human habitation. I notice a lot of stone circles and megalithic tombs nearby, put on my boots, and start walking in search of our past.

~

Holohan's reopened, rather unexpectedly, last night. A young lad, mid-twenties, local, has taken it on. He says he wants to make a real go of it, but not in quite the same way that we've become used to people making a go of things. He tells us he doesn't want to take a wage from it, not for a time at least, and that he just wants the pub to remain open, for there to be a place for people to meet.

He warns Paul and me that whoever loses at chess tonight has to drink a 'cement mixer' – something that, in these less civilised parts, involves knocking down a whiskey, an Irish cream and a half pint of stout in one go – for their troubles. On the house. Neither of us wants to lose at the best of times, but that gives us added incentive, as we both have plenty to do in the morning, none of which will be easier with a hangover. We tell him that if the loser has to drink one of those concoctions then the loser is paying for it too. He's not the only one who wants to see the place stay open.

It appears there are others, too. Along with the regular old boys at the bar, there's young people in tonight. And women. Yes, women. There's a sense of renewal about the place, pub and people both acting like two lovers reunited after a childish spat.

I lose. A man has to keep his word. We tussle over the cash, before I tell him that I won't be back unless he takes it. He looks confused. The hangover's not going to help with the manure collection in the morning, but who cares. There are strong signs of life, and that's all that matters, really.

As I go to leave, the landlord offers me a lift home as he's clearing up. It's on his way, he says. I tell him that I'll need the walk home to clear my head before bed.

It's late when I get home. Not a single car passed me.

~

Publishers love a happy ending. Publishers love a happy ending because readers love a happy ending. Readers love a happy ending because, well, who wouldn't? We all long for happiness in our own lives, and it speaks well of us that we still want this for others too. The only problem is that, viewed through a short-sighted lens, reality doesn't always do happy endings.

I received a letter, this morning, from Kirsty. She's been on the road for the last four weeks, busking, visiting other communities and landscapes, rediscovering herself and tackling her own deepest issues and fears. I usually receive a letter from her every week. It has felt difficult not being able to write back, as she has had no fixed address to respond to, but I love hearing about her adventures.

This morning's letter is different. In it she tells me that she has decided she won't be coming back here to live. She has other paths she feels called to explore, and she wants to give all of her energy to that, on her own.

I read it again. And again. It hits me right in the heart. I'm not usually one to cry, but the tears come down hard. We have been together for three years, and I had assumed we would spend the rest of our lives together. My head is going around in circles, thoughts left with no avenue of expression. As I sit on the windowseat of the cabin I can see her in the patchwork curtains she made, in each grain of wood, in every little detail she perfected.

If I had my time again, I tell myself, I would do things very differently. I would certainly prioritise spending time with her, doing things she loves, over everything else that wasn't absolutely necessary. But life doesn't always give second chances, and you just have to be grateful for the one you got. Today I don't care for wise words like that, though. Today I'm heartbroken.

One thought loops through my head. Why can't I learn. Why can't I fucking learn.

Reality kicks in. Being cut off from a world increasingly connected by fibre optic cables instead of eyes, meeting new people isn't going to be easy. They say there are plenty more fish in the sea, but from where I sit now, with the letter in my hand, all I can see is the pond next to the garden. I stop that trail of thought early, though, before it grabs a hold. No use.

The tears dry up, the cloud passes. It's important to give such emotions their time, but it's also important not to give them a second longer. No point wallowing in self-pity, there's too much of life to be had. And if love is to mean anything, it must mean wanting the best for the person you claim to love, though your own heart feels like it is being ripped from your chest as you let them go. Losing someone you cherish in your life can feel brutal, but it's the risk we gladly accept when we open our hearts to the immensity of love.

I put the letter away. I have beds to manure, wood to bring in and water to collect.

~

I look at the pencil in my hand, and put it down on the table for a moment. It's a hexagonal, machine-cut piece of wood – I'm not sure what species, but I suspect cedar – with what I guess is a thin rod of graphite running through it, finished in yellow, white and blue paint. It looked identical to all the other pencils in the 'H' box when I bought it. The barcode confirms the standardisation is complete.

I buy my pencils from the small art shop in our nearest town, specifically because the owner and his wife are trying to get the money together to jack in the world of business and, instead, move to a smallholding in Connemara. If he were to give me a guided tour of the entire process of making one – from the building of the roads to get the workers from suburbia to the factories, to the extraction and felling of the materials, etc. – I wouldn't want to buy a single pencil. But he doesn't, and I do.

I look at the pencil on the table and wonder if I should ever pick it up again. God knows there are plenty of reasons not to. For one thing, there's the ecological impact of the infrastructure required to make one. Yes, it's a fraction of the embodied energy of a laptop and the World Wide Web, and a pencil can't make you impulse-buy online or take up your morning with distractions like celebrity news, porn or social media. The degrees do matter, but they are still only a matter of degrees.

This sliver of wood also makes my body hurt in ways that lumping logs around all day doesn't. Bad ways. My neck in particular. I'm a physical animal – we're all physical animals – and so the sedentary moments, if too long, don't come easy to me. And while most days I feel fortunate to be able to work both my head and my hands, with each informing the other, writing can often drag me away from the present moment, the place where I physically am, and from what is immediately around me; and I prefer to be where I am, and not somewhere in time or space that I'm not. Some days I question why I bother; after all, the natural world isn't going to rack and ruin for the want of books in the world.

It's on those days that I sometimes close my eyes and imagine myself outside, doing something intrinsically useful like rewilding land, restoring the expansive Great Forest of Aughty that Brian Boru once loved. Now that would be something. The rivers running free of slurry, topsoil and agricultural chemicals, with salmon and brown trout – maybe even sturgeon – returning in numbers that our ancestors would have once taken for granted.

I see fine, healthy populations of ladybirds, craftspeople, starlings, red deer, hedgehogs, musicians, mycorrhizal fungi, Irish mountain hares, goshawks, fishermen, Daubenton's bats, moths, grasshoppers, corncrakes, honeybees, skylarks, witches, wood mice, otters, faeries, hen harriers, bards, pine martens, earthworms, toads, feral goats, badgers, pygmy shrews, curlews, growers, stoats, foxes, lacewings and golden eagles, all adding their own nature to the complexity and wonder of this place. And, if I allow myself to fantasise, a pack of wolves to fit the range. Heady dreams, heady dreams. But a man has to dream, even if he can't afford an acre, let alone 100 square miles of landscape. Maybe I'll find a way.

I pick up the pencil. For all of its faults, at this moment in history the written word – in lieu of a strong oral culture – still offers us an avenue to connect the present with the past, reminding us of perspectives and ways that we have forgotten. Ways that may have value once again in a different kind of world, one that may arrive whether we like it or not; ways that could help us regain our sense of shared humanity and contentment, restore our mental health, and teach us the humility we will need to take our rightful place in the fabric of life again; ways that might even show us the path back home.

I pick the pencil up because, sore necks and unfulfilled arboreal dreams aside, something still implores me to. I suppose I just have to trust that feeling and to keep informing it with the intellect, the heart and the soul. Moving from laptop to pencil was a big step for me, one that I didn't think I would be able to make. But the pencil has enabled me to really start enjoying the process of writing. The writing itself is slower, yet somehow I manage to get more done in less time. The pencil has changed how I think, slowed me down, and made my words human again.

If the feeling to keep picking up the pencil does persist a little longer, I tell myself that next time – when I'm more competent – I'll write entirely with my own pen (quill), ink (ink-cap mushrooms) and paper (birch polypores and dryad's saddle fungus). But one step at a time, I have to remind myself, over and over and over again.

~

Friday, lunchtime and I'm in the hot tub with Edward Abbey. All 336 pages of him. It's almost too hot – almost – so I dangle my legs over the sides into the chilly air outside. Yesterday's warm front (persistent rain) is predictably followed by today's cold front (sunshine sporadically interrupted by showers). One moment the sky above is blue, glorious, unlimited, and I feel blessed for such a spectacular existence; the next, all is dark, ominous, enclosed, and I feel blessed for such a spectacular existence.

Raindrops appear as hundreds of small, clear spears of bath-water shooting skyward. The road to the south sounds unusually busy for this phase of the day and week, and I remember that it must be the last shopping weekend before Christmas. Like the clouds above me, I know it will blow over soon.

As I lie back, gazing towards the heavens, my thoughts unexpectedly turn morbid. As rain pounds my head, refreshing me, I ponder death.

The death of this thing I think of as me. There will come a time, sooner or later, when I'll never walk the beach with my mother and father, will never work and eat and drink with people I care about, will no longer experience the world and its breathtaking web of life through these particular eyes and ears, this particular nose and mouth and skin. A hard thought, but an important and inescapable one.

The death of this place I now call home. One day, hopefully after buzzards have gouged out my eyes, our cabin will also return to the land from which it came, leaving no trace that it ever existed. Perhaps someone will build their own dwelling, in the fashion of their own soul, where ours stands now; or, better still, an oak will become ancient here and provide shelter for a thousand different species. A comforting thought.

The death of all things. The sun will eventually burn itself out, leaving nothing, but not as we know it. It's difficult for my meagre, mediocre mind to grasp the fact that the land under this hot tub will, in a few billion years or so, be no more. Not just that it will be transformed into ocean or desert or glacier, or be populated by creatures I cannot even imagine, but that it will not even exist, gone without a trace of me or them or it.

I notice Packie stutter past in his tractor, white smoke splurting all around, his engine chugging along. He'll be gone one day, too. Yet the fact that one day he is going to die somehow makes it feel more important, not less, to take good care of him between now and that inevitable moment.

~

Standing on the white sandy beach of the Great Blasket again, it's hard not to feel pensive as I look towards the abandoned village above. Stone cabins that once homed laughter, folklore, tears, prayer, song, gossip, feasts, grief, hunger, warmth, tiredness, despair, friendship and hope – the whole gamut of human experience – now sit silent, lonesome, fossils of an extinct people.

But as I turn back towards the wild, indifferent expanse of the Sound, it strikes me that the Great Blasket may now be more inhabited by life than during any other period over the last three hundred years.

Basking in the warm midday sun, a colony of seals – perhaps a couple of hundred strong – are enjoying post-human life on An Trá Bhán (the White Strand), free from the fear of being clubbed to death. As I walk up the hill towards An Dún, I meet a couple of researchers, on their hands and knees, making audio recordings of the Manx shearwater, who nest on the island at night. Storm petrels are breeding in their thousands. Looking at the warrens, I don't suppose the rabbits are mourning the lack of children releasing ferrets down their holes. This uninhabited rock is now home to black guillemots, puffins, seagulls and razorbills, all living in abodes humbler still than small stone cottages. There are probably a thousand other creatures, too, if only I had the eyes to see them. While small human settlements have come and gone, these creatures are still here, stubbornly staying put, living on the cliff edge, making their living from the ocean, or stealing chips from tourists on their own self-propelled trips to Dingle.

As I walk back towards Tomás Ó Criomhthain's old house, it crosses my mind that perhaps this small insignificant outcrop may be better off without human civilisation, at least until we are prepared to learn how to live in dialogue with all things wild once again. And maybe its role for now is a bit like 'the red dust and burnt cliffs and the lonely sky' of Edward Abbey's Utahan desert: a sanctuary of solitude for those willing to brave the place and, through doing so, come face to face with themselves and the rocks and crashing waves and raw elements of life.

With that, my mind drifts back to Knockmoyle. Will it go the same way? Hard to say. Maybe the small farms which comprise it will merge into bigger and bigger mechanised operations run by fewer and fewer people. Or maybe, as small family farms give way under the tyranny of urbanisation, industrialisation and rapacious capitalism, something else will happen. Maybe wildlife, wetlands and woodlands will return to the financially unviable green fields of grass and rush and create thriving local economies for those myriad creatures which we, in our foolishness, forgot and forgot and kept forgetting. Or maybe a miracle will occur, as miracles sometimes do, and the young men and women who moved away will become tired of being farmed in cities, and instead long for something new, something old, something a bit wilder than getting pissed in a chain bar on a Friday night, and in doing so reinvigorate places like this with youth and song and dance.

However things unfold, the wheel of life will keep turning, regardless of whether we keep in tow with it or not. Still, I can't help but feel Knockmoyle wouldn't be the same without Packie.

~

Sometimes, when I catch myself emptying a bucket of my own shit, butchering a deer, shifting manure in the pissing rain, or doing any of the thousand other small things which make up my life – things that, at other times, would have seemed hare-brained, unethical, absurd – a feeling of 'how the hell did I get here?' comes over me. *This* was never part of the programme. Like everyone, I had dreams of success and the good life, but somewhere along the track, a place I can't quite put my finger on, the definition of those words began to change, and my life with it.

People ask me all the time if I will continue to live in this way for the rest of my days. No more than the next person, I cannot see what the future holds. But after ten years of letting go of the trappings of modern life, I feel like I have only just begun to scratch the surface. There are depths to the human experience that I still can't even imagine, buried as they are under the layers of ambition, plastic and comfort that we've all been cloaked in, through no fault of our own, from the moment we were born. If anything, I want to explore these depths further, to see what treasures hide below. I certainly have no longing, as I write these words, to fall back into a way of being which sells comfort for the price of everything it is to be human.

If it's possible – and I'm not convinced at all that it is – I want to take off the manufactured lens of industrial civilisation and see the world through my own eyes, on its own terms. At our most fundamental level we're animals, yet I've still very little idea of what that really means. Many years ago, I decided that instead of spending my life making a living, I wanted to make living my life. That feels as true to me today as it did then. For as Patrick Kavanagh said, I no longer wish to hawk my horse, my soul, to the highest bidder. I've tasted the grass 'on the south side of ditches', and I've found it to be sweeter than that on the farms.

At a talk I gave a few weeks ago, someone asked me about what I will do when I get old. I said that, like everyone, I will die. I have no desire to be the man who made it safely to death, wearing an oxygen mask at eighty-eight, afraid of letting go, terrified of what might come next. Our relationship with death profoundly changes our relationship with life. It's all too easy to live a long, unhealthy life without having ever felt truly alive.

What will happen between this present moment and having my bones licked clean by a ravenous wild animal? As I say, I don't know, as I'm no longer blessed with the certainties of youth. The more I explore, the less I seem to know, and I'm starting to like it that way. If someone comes along and convinces me that all of the impedimenta of contemporary society – the screens, the engines, the switches – are actually life-enriching, life-affirming, life-giving, then I'll change tack and start sailing towards that shore, to see if they're onto something. But for now, I'm going to try to stay in the only place that makes sense to me: the bloody, sublime, mucky, sweaty, breathtaking world of life.

What I can say, right now, is that I've fifty willow cuttings to put in the ground, and that if I don't want to feel hungry tomorrow, I'd do well to get off to the river, to see if I cannot better understand it.

The Complexities of Simplicity

My wish simply is to live my life as fully as I can. In both our work and our leisure, I think, we should be so employed. And in our time this means that we must save ourselves from the products that we are asked to buy in order, ultimately, to replace ourselves.

Wendell Berry, *The Art of the Commonplace* (2002)

Before I go, I should say a few words about one confusing little word: simple.

My ways are often described, sometimes by myself, as 'the simple life'. Interpreted one way, it is entirely misleading, as my life – my livelihood – is far from simple. It is actually quite complex, but is made up of a thousand small, simple things. In comparison, my old life in the city was quite simple, but was made up of a thousand small, complex things. The innumerable technologies of industrial civilisation are now so complex they make the lives of ordinary people simple.

Too simple. I, for one, got bored doing the same thing day in, day out, using complex technologies which I suspected made those who manufactured them bored too. That's partially why I rejected them. With all the switches, buttons, websites, vehicles, devices, entertainment, apps, power tools, gizmos, service providers, comforts, conveniences and necessities surrounding me, I found there was almost nothing left for me to do for myself; except, that is, the one thing that earned me the money to acquire all the other things. So, as Kirkpatrick Sale wrote in *Human Scale*, my wish became 'to complexify, not simplify'.

But while the technologies and bureaucracies of industrial society are complex, the society itself isn't. The agricultural scientist Kenneth Dahlberg described it best when he said: 'Industrial societies are "complicated" (like a clock which has many interlocking parts but only a few "species" – gears, springs, bearings etc.) . . . [They] are not complex.' Sale added in *Human Scale* that: 'It is our modern economy that is simple: whole nations given over to a single crop, cities to a single industry, farms to a single culture, factories to a single product, people to a single job, jobs to a single motion, motion to a single purpose.' Compared to industrial civilisation, a thriving wilderness is remarkably complex.

Interpreted another way, there is a timeless simplicity about my life. I have found that, when you peel off the plastic that industrial society vacuum-packs around you, what remains – your real needs – could not be simpler. Fresh air. Clean water. Real food. Companionship. Warmth, earned from wood chopped with your own hands using a convivial tool whose only input is care. There's no extravagance, no clutter, no unnecessary complications. Nothing to buy, nothing to be. No frills, no bills. Only the raw ingredients of life, to be dealt with immediately and directly, with no middlemen to complicate and confuse the matter.

Simple. But complex.

All of this talk of complexity and simplicity is starting to hurt my head. I notice the chickens have put themselves in and are waiting for me to close the door – like all domesticates, they've grown used to being looked after by somebody else – and I've a few of those other thousand things to get on top of too, so I had best get on.

~

Kirsty has come to visit. It's only a flying visit, but I intend to savour the moments life grants me from here on. My love for her has not diminished one iota, even if its expression is now taking another form. Our friendship will never be in question. Having her here has emphasised how any way of life, no matter how rewarding, can feel lacking without love, without that daily reminder to recognise the beauty in everything and everyone around you.

As I am about to put my pencil down, perhaps for a long time, I notice her across the cabin, crunching up dried herbs, storing them, making labels for each one individually, all in her own fashion. She is so absorbed in it that she has no idea that I am watching her. The complexities of simplicity haven't always been easy for her, just as they will not always be for anyone from our generation who grew up with too much comfort and convenience. I know this too well myself. But with time, the complexities do become beautiful, and she has become a more complex person in the most beautiful sense, even though I will no longer be the one to appreciate it on a daily basis. She's a natural herbalist; natural, because she loves it.

She finally notices me watching her, and smiles. I could want for nothing more than what I have in this moment. Its brevity is irrelevant. I go over to her, and when I've got her off guard I tickle her under the arms until she is lying on the floor and laughing in that uncontrollable way that I know no other person in the world laughs. When she can't take it any more I let her up, and we go about storing selfheal, yarrow and silverweed together. Those other jobs can wait until tomorrow.

Postscript

A couple of months after finishing this book, every word of which was written and rewritten by hand, it started to become increasingly clear to me that, if it were ever to see the light of day, it was going to have to be typed up. Publishers, my agent, even a few friends – they were all saying the same thing. Throughout the writing process, I had always held out faint hope that there might be another way, that these words could be presented in the handcrafted form of the person who had thought them. But deep down I always knew that, for various understandable reasons, they were going to have to be turned into the easily readable, easily editable, easily publishable form of type.

Still, I stubbornly resisted for a time. That was until I realised that the situation was simply asking another variation on the same question that has followed me since I gave up money a decade earlier. From the moment I decided to reject much of what civilisation has to offer, I've had two options. One was to 'go native', as they say, and let industrial society go to hell (where, as I understand it, it seems hell-bent on going anyway). The main criticism of this approach has always been that it's a form of selfish escapism, and that in doing so you're helping no one but yourself. Though I disagree with both the logic and practical experience of such criticism, it has never been a course of action that I've felt comfortable with myself.

The other option has been to live the life I want to lead, but to do so as part of the society whose ways I seek to question. The main criticism of this approach is hypocrisy, as it inevitably means engaging in the ways of that same society, ways which are often diametrically opposed to your own. Yet there is something about this approach, something beyond the kind of hardline ideology that I'm all too often prone to, that feels true to me. Even so, I've never really understood why hypocrisy has such a bad reputation anyway. As David Fleming wrote in *Lean Logic*, 'There is no reason why he [the hypocrite] should not argue for standards better than he manages to achieve in his own life; in fact, it would be worrying if his ideals were *not* better than the way he lives.'

As you are reading this, it's obvious that I chose the hypocrite approach. Once I agreed to have it typed up, my options forked off again. It became a choice between getting someone else to type it up for me – Wendell Berry's wife types up his words on a Royal Standard typewriter, bought in 1956 – or to type it up myself. As I knew no one for whom such a task would have been a labour of love, I took the difficult decision to make a one-off, as-brief-as-possible exception to my tech-free life. I typed it up myself.

I had, and still have, big reservations about that decision. I wonder if there really was no other way, if the decision was coming from the right place, or if I should have just binned the manuscript and taken the first option instead. Either way, I won't try to defend the decision. It was a conscious choice, weighed up, and in the end made for reasons that I felt were more important than trying to be right or some misplaced notion of ideological purity. Walt Whitman understood this sentiment when, in his poem 'Song of Myself', he wrote:

Do I contradict myself?

Very well then I contradict myself,

(I am large, I contain multitudes).

As I said in the prologue, which was hand-written long before this postscript, these days I prefer to explore the complexities of both living and communicating this older way of life, in the modern world, than to pretend that the situation is always black-and-white. It's not. It never has been.

Aside from the philosophical aspects, the decision had practical implications. I wanted to get it over as quickly as possible, so that I could return to the life I had come to love. So, having not sat in front of a screen for eighteen months, I suddenly found myself spending twelve hours a day, for seven days, turning my carefully hand-written manuscript into an electronic, publishing-friendly, generic form. Going back to using a computer – even for this relatively brief, defined task – was almost as insightful an experience as giving it up had been in the first place.

To begin with, I could barely type. Qwerty no longer made sense to me as I searched for the initial letter of the title page. When I worked in business it was normal for me to spend days on end on my arse in front of a screen. But now my head felt so fried by lunchtime each day that I had to restrain myself from opening a bottle of wine. I was less able to cope with something most of us now consider to be run-of-the-mill, and I couldn't work out if that was a sign of mental weakness or strength (Krishnamurti once remarked that 'it is no measure of health to be well adjusted to a profoundly sick society'). By the end of the first day, my back ached, and the repetitive strain injury I had carried for years on my right wrist had made a mild and temporary comeback.

But the effects ran deeper. I felt less purposeful, like I no longer knew what my life was about, or what I stood for. By evening I felt entirely disconnected from the landscape around me, like I was no longer a part of it, but some strange virtual universe instead. The natural light hurt my eyes as I re-emerged outside.

In some ways it was good and important for me to temporarily re-enter that world of things, so as to dispel any romantic memories I had about life being much better and easier with machines. The experience of it was such that, having made the compromise, I'm not sure I would make it again.

A couple of days after I finished typing I slowly felt the effects of screen-staring and the sedentary life wear off. I found my connection to my place return, like I belonged here again. Still, I would do well to remember how it felt for those seven days when I'm next out hand-washing the clothes.

That's enough typing from me. Outside is calling. I can hear a magpie squawking madly as another is under a tree, plucking out the feathers of a flapping wood pigeon. Up above a bullfinch sings a duet with his mate, for love, or life, or the love of life.

Outside. That's where I'm meant to be.

A Short Note on the Free Hostel

As mentioned throughout the book, on our smallholding we run a free hostel, event space and *síbín* called The Happy Pig. People come to stay in it for all sorts of reasons: some to get stuck in and experience a more elemental way of life; others to take time out to read, walk, play music or be creative; many seem to be trying to work out what they want to meaningfully do with what the poet Mary Oliver calls their 'one wild and precious life'. It hosts occasional courses, evening events and shindigs, and it is available to groups to use for free. I could write a book (but I won't) on many of the characters who have passed through here.

It works a bit like a bothy. Everyone is welcome. You can stay for up to three nights, and as long as you're not bothering anyone you can stay for longer too. Some people have stayed for months. We have no website, and as such we are like the hostels of old. In fact, as such it is like the everything of old.

We don't give out directions. You have to follow your nose, and your own innate sense of adventure. Knock on doors, ask shopkeepers, take wrong turns. Don't even think about using your smartphone. We have no booking system. You can write to me in advance – to which I'll unlikely reply, unless you want to organise an event – or you can do as most do and just show up at the door. I'm sure you'll figure it out. While we accept donations, none are expected, and anything that is contributed towards the building's small costs, by those who can and want to, is done strictly anonymously.

A few important notes. If you come from overseas, we ask that instead of flying you come by land and sea. Better still, walk or hitch or crawl on your hands and knees. Anything other than getting here effortlessly. When you do get here, we ask that you are self-reliant for food and your own entertainment, though you are more than welcome to whatever we can offer of both. If you have a musical instrument or a song, bring it. Enthusiasm is always welcome. Those of us who live here permanently don't always have the time or inclination to hang out, but we often do.

So the guidelines are quite simple. Enjoy your time here. Be mindful of what you use and why you're using it. And leave the place at least as well as you found it. A bit like life, really.

Select Bibliography

Abbey, Edward, *Desert Solitaire: A Season in the Wilderness* (McGraw-Hill, 1968)

——*The Journey Home: Some Words in Defense of the American West* (Plume, 1991)

Ansell, Neil, *Deep Country: Five Years in the Welsh Hills* (Penguin, 2012)

Berry, Wendell, *The Peace of Wild Things* (Penguin, 2018)

——*The World-Ending Fire: The Essential Wendell Berry* (Allen Lane, 2017)

Cahalan, James M., *Edward Abbey: A Life* (University of Arizona Press, 2001)

Carney, Michael and Hayes, Gerald, *From the Great Blasket to America: The Last Memoir by an Islander* (The Collins Press, 2013)

Colvile, Robert, *The Great Acceleration: How the World is Getting Faster, Faster* (Bloomsbury, 2017)

Connell, John, *The Cow Book: The Story of Life on a Family Farm* (Granta, 2018)

Deakin, Roger, *Wildwood: A Journey Through Trees* (Penguin, 2007)

Diamond, Jared, *Guns, Germs, and Steel: The Fates of Human Societies* (Norton, 1999)

Dillard, Annie, *Pilgrim at Tinker Creek* (Harper Perennial Modern Classics, 2013)

Emerson, Ralph Waldo, *The Essential Writings of Ralph Waldo Emerson* (Modern Library, 2000)

Finkel, Michael, *The Stranger in the Woods: The Extraordinary Story of the Last True Hermit* (Simon & Schuster, 2017)

Fleming, David, *Lean Logic: A Dictionary for the Future and How to Survive It* (Chelsea Green, 2016)

Fogarty, Pádraic, *Whittled Away: Ireland's Vanishing Nature* (The Collins Press, 2017)

Griffiths, Jay, *Pip Pip: A Sideways Look at Time* (Flamingo, 1999)

Hayes, Gerald W. and Kane, Eliza, *The Last Blasket King: Pádraig Ó Catháin, An Rí* (The Collins Press, 2015)

Jones, Tobias, *A Place of Refuge: An Experiment in Communal Living, The Story of Windsor Hill Wood* (Riverrun, 2016)

Kavanagh, Patrick, *A Poet's Country: Selected Prose* (The Lilliput Press, 2003)

——*Collected Poems* (Allen Lane, 2004)

Kelly, Kevin, *What Technology Wants* (Penguin, 2011)

Kingsnorth, Paul, *Confessions of a Recovering Environmentalist* (Faber & Faber, 2017)

Langlands, Alexander, *Cræft: How Traditional Crafts Are About More Than Just Making* (Faber & Faber, 2017)

Lawrence, D.H., *Lady Chatterley's Lover* (Penguin, 1960)

Leopold, Aldo, *A Sand County Almanac* (Oxford University Press, 1949)

Lopez, Barry, *Arctic Dreams: Imagination and Desire in a Northern Landscape* (Bantam, 1987)

Mac Coitir, Niall, *Ireland's Wild Plants: Myths, Legends and Folklore* (The Collins Press, 2015)

——*Ireland's Birds: Myths, Legends and Folklore* (The Collins Press, 2017)

Macfarlane, Robert, *Landmarks* (Penguin, 2016)

——*The Old Ways: A Journey on Foot* (Penguin, 2013)

Mears, Ray, *Essential Bushcraft* (Hodder and Stoughton, 2003)

——*Outdoor Survival Handbook* (Ebury Press, 2001)

Michelet, Madame, *Nature: or, the Poetry of Earth and Sea* (T. Nelson and Sons, 1880)

Monbiot, George, *Feral: Rewilding the Land, Sea and Human Life* (Penguin, 2014)

Muir, John, *Wilderness Essays* (Gibbs Smith, 2015)

Mumford, Lewis, *The Myth of the Machine: Technics and Human Development* (Harcourt, Brace & World, 1967)

——*The Myth of the Machine Volume II: The Pentagon of Power* (Harcourt, Brace & Jovanovich, 1970)

Mytting, Lars, *Norwegian Wood: Chopping, Stacking and Drying Wood the Scandinavian Way* (MacLehose Press, 2015)

Nearing, Helen and Nearing, Scott, *Living The Good Life: How to Live Sanely and Simply in a Troubled World* (Schocken, 1989)

Ó Catháin, Gearóid Cheaist, *The Loneliest Boy in the World: The Last Child of the Great Blasket Island* (The Collins Press, 2015)

O'Connell, Mark, *To Be a Machine: Adventures Among Cyborgs, Utopians, Hackers, and the Futurists Solving the Modest Problem of Death* (Granta, 2017)

Ó Criomhthain, Tomás, *Island Cross-Talk: Pages from a Blasket Island Diary* (Oxford University Press, 1986)

——*The Islandman* (Oxford University Press, 1951)

Ó Guithín, Micheál, *A Pity Youth Does Not Last: Reminiscences of the Last Blasket Island Poet* (Oxford University Press, 1982)

Ó Súilleabháin, Muiris, *Twenty Years A-Growing* (Oxford University Press, 1933)

Rebanks, James, *The Shepherd's Life: A Tale of the Lake District* (Penguin, 2016)

Richards, Matt, *Deerskins into Buckskins: How to Tan with Natural Materials* (Backcountry Publishing, 1997)

Sale, Kirkpatrick, *Human Scale* (Martin Secker & Warburg, 1980)

Sayers, Peig, *Peig* (The Talbot Press, 1983)

Snyder, Gary, *The Practice of the Wild* (Counterpoint, 2010)

Thoreau, Henry David, *The Journal 1837–1861* (New York Review of Books, 2009)

——*The Portable Thoreau* (Penguin, 2012)

Tree, Isabella, *Wilding: The Return of Nature to a British Farm* (Picador, 2018)

Whitman, Walt, *The Works of Walt Whitman* (Wordsworth Editions, 1995)

Wiseman, John 'Lofty', *SAS Survival Handbook* (Collins, 2003)

Wohlleben, Peter, *The Hidden Life of Trees: What They Feel, How They Communicate – Discoveries from a Secret World* (William Collins, 2017)

Acknowledgements

It's quite a complicated thing writing the acknowledgements for a book when you live as I do. It is difficult to know who or what to include, or leave out, when everyone and everything you have encountered in your life has influenced it, and the writing which emerges from it, in both subtle and profound ways.

And so I will keep it simple: thank you, Creation.